Energy Derivatives: Trading Emerging Markets

Energy Derivatives: Trading Emerging Markets

By Peter C. Fusaro and Jeremy Wilcox

Energy Publishing Enterprises
New York, New York

Library of Congress Cataloging-in-Publication Data

Fusaro, Peter C. and Wilcox, Jeremy
Energy derivatives: trading emerging markets/Peter C. Fusaro and Jeremy Wilcox
ISBN 0-9745391-2-0-
1. Energy industries. 2. E-commerce. 3. Energy trading.
Library of Congress Card Number: 00-106540

ENERGY PUBLISHING ENTERPRISES

Energy Publishing Enterprises
A Division of Global Change Associates Inc.

The editors for this book were *Peter C. Fusaro and Jeremy Wilcox.* It was set in 11 point Palatino and designed by *S. Kloos Communications Inc.*

Additional books can be ordered directly from Whitehurst and Clark Book Fulfillment Inc., Raritan Industrial Park, 100 Newfield Avenue, Edison, NJ 08837; (Fax) 732-225-1562 (use order form at end of book).

Energy Publishing Enterprises books are available at special quantity discounts to use for promotional, educational or corporate training programs. For more information, please write Energy Publishing Enterprises, 2124 Broadway, PMB #195, New York, NY 10012, USA. Or contact your local bookstore.

This book is dedicated to our children: Laura, Sebastian and Barnaby

CONTENTS

CONTRIBUTORS

Editors:
Peter C. Fusaro and Jeremy Wilcox, principals in Global Change Associates

Jim Banks
Freelance Energy Writer

George Campbell
Manager, Power Marketing
SCANA Energy Marketing

Michael Coleman
Managing Director
FSD International

Louise Croucher and Alan Gillepsie
Tradition Financial Services

Garth Edward and Matthew Varilek
Natsource Tullett & Tokyo

Dr. Antoine Eustache,
Global Index Editor
Dow Jones Newswires

Andrew Feachem
Weather Derivatives
Eurobrokers

Lin Franks and Terrence M. McGee
Andersen Consulting LLC

Ms. Patricia Hemsworth,
Director, Natural Gas, Electricity
& Coal Marketing
New York Mercantile Exchange

Seana Lanigan
Formerly of the International
Petroleum Exchange

Michael Moore
Managing Director
Amerex Bandwidth

CHAPTER 1

New Market Developments

By Peter C. Fusaro and Jeremy Wilcox

The process of energy reform, which started with the ending of Official Selling Programs in the 1970's, is now nearing its completion. Catalyzed by the global deregulation and liberalization of electricity markets, and facilitated by the evolution of electronic trading, the energy markets are now more competitive than at any time in their history.

Competition, which will reach its threshold within the next ten years, is already being supplemented by the market dynamics of consolidation, which in turn will lead to convergence across all energy and related markets. The end vision is one of a small number of international energy corporations with multi-energy commodity product verticals.

The development of convergence is not restricted to the 'pure' energy markets of oil, coal, gas and electricity but also encompasses the ancillary energy markets of weather, emissions and bandwidth. As oil markets continue to mature these ancillary markets, plus electricity and gas, will exhibit dynamic growth as the time frame for the effective commoditization of new markets reduces almost exponentially. They will also mature much more rapidly than in the past.

Of these new ancillary markets the most dynamic growth is expected in bandwidth. Companies such as Enron have dedicated huge resources in the telecommunications sector, which is forecast to exceed the current market capitalization of energy within three years. Similarly the value of emissions cannot be underestimated. While the Kyoto Protocol is unlikely to be ratified in its current format, the expectation is that a 'Kyoto II' will emerge in the next few years. Led by the North American market, emission trading is evolving as the only proactive approach to reduce greenhouse gas emissions. In Europe, emission trading programs are being developed in Scandinavia, Germany, France and the UK.

With the high volatility associated with electricity the need to hedge volume risk as well as price risk in order to prevent outages has promoted the trading of weather derivatives. Again. Europe and the rest of the world is following the lead of the US markets in weather hedging and risk management.

In tandem with the structural reform of the energy markets there has been a gradual but prominent move in the trading infrastructure. Futures markets, once the bastion of trading, have lost their luster. No longer the innovator of trading contracts and strategies, futures markets have given way to Over-the-Counter (OTC) markets. New markets in electricity, coal, weather, emissions and bandwidth are being developed in the OTC markets. Although still compatible, the value of futures markets is being more and more restricted to the front end of the forward curve.

Futures markets now have to address not only the competition provided by the OTC markets but also the development of electronic OTC trading platforms. Prominent financial exchanges such as Deutsche Borse and London International Financial Futures Exchange (Liffe) have signalled the way forward by spinning off dot.com companies to broaden their appeal and open up new market opportunities. Traditional energy exchanges will follow suit. New York Mercantile Exchange has already outlined its move into the electronic arena, and the others will follow.

The development of electronic and Internet trading platforms is a particular threat to the historical franchise business of energy exchanges. Through globalization and convergence the provision of trading instruments is not limited to a particular market sector. Financial exchanges are diversifying into energy and commodities and energy exchanges have to adopt a similar approach. SwapNet, the Internet-based energy swaps transaction platform, has recently broadened its product verticals into bonds. There is no reason why energy exchanges, whether they be futures or OTC based could not provide non-energy contracts.

The value of the Internet is that it is a global marketplace with no borders or barriers, and as such provides the ideal medium for so called 'one-stop-shop' platforms—energy, commodities and financials.

The new developments in the energy market are being driven by the reform and restructuring of the underlying physical markets. As a result of competition, and in particular retail competition, the value of derivative applications and risk management strategies increases in importance. And with the emergence of the ancillary markets of emissions and coal new bundled transactions are emerging. Coal is being traded with bundled sulfur emissions to ensure the level of sulphur in the coal is properly evaluated. Temperature-related electricity swaps and options are being transacted to provide combined price and volume hedges. This convergence of energy and energy-related markets will lead to the creation of a true BTU energy market.

This book is the second primer on energy market risk management developments. It is meant as a companion piece to **Energy Risk Management** (Fusaro, McGraw-Hill, 1998). It is envisioned that a third book will be published next year dealing with the emerging markets of petrochemical hedging, LNG hedging, newer electronic platforms, and other market developments. The present book will take the reader through the latest market developments and examine those developments in layman's terms. The book is meant as an educational vehicle for learning by making complexity simple and easy to understand for anyone interested in the subject of energy risk management. The initial chapters of this book provide the fundamentals on energy trading for electricity, natural gas, emissions, weather, coal and bandwidth. The newer platforms of electronic trading, index construction, risk management software systems, and retail markets are brought forth in the second part of this text.

The book is organized in the following manner. Chapter 2 examines the developments in electricity trading markets in both Europe and European markets which is provided by Patricia Hemsworth of the New York Mercantile Exchange. Chapter 3 provides insights from Andrew Feachem of Eurobrokers into the weather derivatives markets which is attracting attention in both the US and Europe. Chapter 4 gives a primer on the evolution of the emissions trading markets and provides insights into its future developments from OTC emissions brokers Garth Edward and Matthew Varilek of Natsouce Tullet & Tokyo.

Bandwidth trading is covered in next two chapters since this is such a new market where the expertise of energy traders is converging with the telecommunications industry. Lin Franks and Terrence Gee of Andersen Consulting explain the fundamentals necessary for market development in chapter 5. Michael Moore of Amerex Bandwidth provides some insights into to its commodity market development in Chapter 6.

Seana Lanigan formerly of the International Petroleum Exchange explains the evolution of European natural gas markets in Chapter 7. Coal trading is explained by Louise Croucher and Alan Gillepsie of OTC brokers Traditional Financial Services (TFS) in Chapter 8.

Co-authors Peter Fusaro and Jeremy Wilcox give the context and lay out the latest technological developments for electronic energy trading in Chapter 9. This chapter examines not only the platforms, but gives some perspective on the evolution of e-commerce solutions for energy trading.

Antoine Eustache of Dow Jones Newswires, and the world's leading expert on index construction for energy brings his unique experience and insights on creating market liquidity for indexes in Chapter 10. Dr. Eustache created the COB index in the US as well as the SWEP and CEPI indexes for European market trading. His latest research is oriented toward bandwidth trading indexes.

Another emerging market, which involves retail competition, is explained by George Campbell of SCANA power marketing in Chapter 11. Retail electric power markets promise to be the ultimate battleground for competitive for energy service providers.

The next two chapters, 12 & 13, focus on energy risk management trading systems. Here, we are fortunate to have a look at end-to-end trading solutions by energy freelancer Jim Banks, and a good look at the architecture for flexible trading provided by industry leader Michael Coleman of FSD International.

Finally, the co-authors once again provide a glimpse of the future of energy trading as the markets begin their ultimate convergence on a multi-commodity platform in the concluding Chapter 14.

Summer 2000
Peter C. Fusaro and Jeremy Wilcox

CHAPTER 2

Electricity Trading: Europe and North America

By Patricia Hemsworth

Over the past ten years, the emergence of a global energy marketplace has prompted governments to pursue free market policies in the electric power sector, designed to promote efficiency, lower prices and stimulate technological innovation. A combination of privatization and deregulation has provided the backdrop for the global restructuring of vertically integrated monopoly utility structures. This process has unfolded differently in various countries as each attempt to address the necessary regulatory changes to restructure the ownership and operation of the electric power sector. Common to all efforts in creating competitive markets is the unbundling of generation, transmission and distribution services in providing non-discriminatory access to the transmission grid. Also important is the decision on whether or how to compensate utilities for stranded assets incurred as a result of the transition to competitive markets. This regulatory reform process, well under way, has stimulated the creation of competitive wholesale and retail markets. As these markets evolve, new secondary or derivative markets are being created to manage financial risk associated with price volatility.

FIRST EFFORTS AT DEREGULATION: UK AND SCANDINAVIAN MARKETS

The first country to institute comprehensive measures to deregulate the electric power markets was the UK. In 1989, it passed the Electricity Act, which outlined a process for restructuring the electric power sector. As part of a broader, ambitious program of national privatizations initiated under then Prime Minister Margaret Thatcher, this "experiment" came to be viewed by the rest of the world as a kind of case study for power deregulation.

The Electricity Act created two fossil fuel generating companies—PowerGen and National Power—and twelve regional electricity companies (RECs). At privatization, the nuclear generating capacity remained under government ownership, but this too was subsequently privatized. Transmission, through the National Grid Company, was owned by each of the RECs.

To balance electricity supply and demand, the UK government set up a PoolCo (the Electricity Pool of England and Wales) as a market clearing mechanism between generators and the RECs. Pool prices were set by generators bidding prices and capacity each half-hour period for the following day, which were then matched to demand forecasts.

The problem with this system is that it is an obligatory pool. It does not allow for bilateral trade between generators and supply companies, and prices are not set by the fundamentals of supply and demand as suppliers have no impact on price setting. However, this pool mechanism or variants of this mechanism have become a model for market-based pricing mechanisms in many parts of the world, including Spain, Australia, Argentina and Asia. A PoolCo model was adopted later in the U.S. by the state of California by the state-mandated PoolCo, California Power Exchange.

To develop trading the UK market introduced Electricity Forward Agreements (EFAs), a secondary derivative market developed to hedge risk associated with market volatility, which could be very high. EFAs are traded in four-hour blocks and essentially cash settled short-term swaps.

In 1999, after a long process of consultation, the UK regulator (Ofgem) announced that the Electricity Pool would be replaced by the New Electricity Trading Arrangements (NETA) in October 2000, which would introduce bilateral trading and provide for futures and forwards markets. The value of NETA is that it closely mirrors the successfully traded gas market, which like electricity is 100% competitive in the UK market. As a result of NETA, a number of futures and OTC platforms have been proposed for the UK market, the first of which (UK Electricity Exchange) launched in June 2000.

In contrast to the UK, Scandinavian countries had had power trading arrangements in place since the 1960s with the creation of Nordel, an organization created to promote cooperation among the largest electricity generators in each country. Power trading arrangements were instituted to maximize efficiency and cost savings. All swaps were based on cost-based pricing, however. Norway began to liberalize markets in 1991 when it allowed their party open access to transmission and distribution networks. Sweden also instituted a market-oriented structure for trading the Scandinavian power markets in 1995, transitioning from a market structure in which almost half of the generation was owned by the state.

In 1996, Nord Pool, an integrated clearinghouse exchange for physical power as well as futures and options on futures was created. It is jointly owned by the governments of Norway and Sweden and features a spot or next day electricity market, a forward market and a futures market. Forty different futures and options products trade on Nord Pool, comprised of different time configurations: futures contracts, which are traded as financial day contracts, weekly contracts four to seven weeks ahead, as blocks of four weeks up to fifty-two weeks ahead, or as seasons up to three years ahead. The futures contract design is distinctive in that the larger time blocks break down into smaller contracts, approaching spot delivery. For example, a four-week block can break down into weekly contracts and weekly into daily. Futures are settled financially. Both Asian and European Options are traded on Nord Pool.

In 1996, Finland allowed open access for buyers purchasing 500 kWh and above. In the same year, Finland launched an integrated spot and futures market, the El-Ex market, which is based on the Nord Pool structure, except that there are season contracts for summer and winter, four-week blocks and hourly contracts, broken down further into day and night contracts. In 1997, it consolidated two grid transmission companies into one, Fingrid.

All Nordic countries have approached the transmission as a natural monopoly. All grid structures are centralized, managed by independent operators and are owned by the respective governments. Similarly, the UK approaches the transmission function as a natural monopoly with the grid being centrally operated and owned by the 12 Regional Electricity Companies.

ELECTRICITY MARKET DEREGULATION IN THE UNITED STATES

In contrast to the power sector restructuring in the UK and Nordic countries, deregulation of the power markets in the United States has been an enormous project, due to the size and diversity of the nation's electric power sector, the complexity of its transmission grid and the fragmented nature of governmental jurisdiction. The diversity of market participants, which include Investor Owned Utilities (IOUs), Federally-owned Power Marketing Administrations (PMAs), municipal power companies, independent power producers and Rural Coops, all with different legacy jurisdictional arrangements also complicates the process as well.

Two key pieces of legislation passed in the 1930s created the monopoly structure of the electric power industry: PUCHA (Public Utilities Control Act) of 1935 and the Federal Power Act. These statutes established the framework for the regulatory compact in the US. The regulatory compact can be defined as the ownership and operational arrangement whereby utilities are rewarded with a guaranteed customer base and fixed rates of return in return for the obligation to reliably serve all within a given territory. This regulatory structure was unchanged for about 50 years and resulted in a configuration of 147 control areas throughout the United States. Like medieval towns, they were enclosed, local and largely self-sufficient. Local utility commissions handled retail and wholesale intrastate pricing as well as many financial and operational issues; federal authorities regulated interstate commerce, wholesale pricing, interconnections and financial transactions. These laws gave the greatest power to state regulatory bodies with federal authority assuming authority for what was not regulated by the states.

Public Utility Regulatory Policies Act (PURPA), passed in 1978, was the next major piece of energy legislation to affect the deregulation of electricity. After the oil shocks of the 1970s, national policy was directed by concerns over energy security. PURPA essentially deregulated the generation side of the business: anyone with the necessary capital was free to build new generation, thereby increasing supply. At this time, new generation technologies were discovered such as combined-cycle technology, whereby hot gases emitted as a by-product of industrial processes or of power generation could be used to generate power. The Fuel Use Act, passed in the same year, allowed for the development of new sources of generation from QFs or qualifying facilities. These generators were generally smaller, more efficient and able to take advantages of new technologies. At that time, regulated utilities were prohibited from using natural gas (due to fears of supply shortages) in new generation technologies, which helped to establish the new QF facilities as lower cost generators. By the time the law was ended nine years later, the new QFs had gained an important foothold in the generation markets, providing new sources of much cheaper power. QFs are non-utility power producers that often generate electricity using renewable and alternative resources, such as hydro, wind, solar, geothermal or biomass (solid waste). QFs must meet certain operating, efficiency, and fuel-use standards set forth by Federal Energy Regulatory Commission (FERC). If a QF meets these FERC standards, utilities must buy or wheel the QF-generated power. QFs usually have long-term contracts for utilities' purchase of this power. Cogenerators may also be QFs. Cogenerators use the waste heat created by one process, for example during manufacturing, to produce steam that is used in turn to spin a turbine and generate electricity.

PURPA became the chink in the wall of the electricity power monopoly system. The QFs fell outside of the jurisdiction of the states and its cheaper power challenged cost based rates. Before PURPA, interstate rates were being set by the FERC at cost-based rates. In the 1980s, Non-Utility Generators (or NUGs) proliferated, creating a new class of Independent power producers and cogenerators. The FERC began to approve these the wholesale power transactions based on market rates. This limited shift toward market-based rates gave new clout to small generators and new options to industrial users, who were able to produce power through cogeneration facilities. It also helped to strengthen the regulatory authority of FERC.

The next major piece of legislation, EPAct, the "Energy Policy Act" passed in 1992, broke down more barriers for a competitive merchant generation market. It created yet another class of generators, the Exempt Wholesale Generators, not subject to state regulation, which were able to produce electricity at market-based rates. These generators were not limited in size or in the type of fuel they could use for power generation, unlike QFs. The legislation assured these generators access to the grid for wholesale bulk power marketing. It gave the FERC the power to require utilities to provide open access to customers. This rule had the effect of accelerating the process of creating an active wholesale marketplace by allowing for more non-utility generation to be built and assuring them access to the transmission grid to sell their wholesale power. Although permission to use another company's grid to wheel or transmit this wholesale power on the grid was voluntary for the transmission owning entity, FERC had the power to mandate open access. These laws provided a structure for market-based bulk wholesale transactions wheeled over an IOU's grid, paving the way for competitive wholesale markets.

While EPAct gave the FERC the power to order open access, the 1996 FERC Orders 888 and 889 went further in devising a comprehensive plan for open access. These rulings required that utilities functionally unbundle the generation, transmission and distribution functions of their power operations, although it did not order divestiture. It leveled the playing field by ordering non-discriminatory ACCESS for wholesale marketers of power to the nation's grid. It also required that utilities post non-discriminatory transmission tariffs on OASIS (Open Access Same Time Information Systems), which became the first major commercial use of the Internet. This ruling also recommended the formation of ISOs or Independent System Operators, to take the place of the utility control operators, who were traditionally employed by vertically integrated corporations. FERC 888 and 889 opened the doors for a robust wholesale trade in the U.S. The movement to assure open non-discriminatory tariffs encouraged the creation of Power Marketers of which there are now over 500 in the United States.

TRANSMISSION GRID MANAGEMENT AND MARKET DEVELOPMENT: FERC 2000

Since, the FERC issued Rule 888 and 889, the number of wholesale transactions in the US increased dramatically. Wholesale power transactions are being done with power being transmitted in unique and different contract paths. In contrast to the simple power swapping arrangements between neighboring utilities in the past, competitive markets were creating new challenges in the operations of the transmission grid, which was not originally built for trading but for regional reliability.

What became increasingly clear to all was that operational inconsistency and lack of coordination in the new world of competitive trading eroded former systems for interregional cooperation regarding reliability. Inadequate methods for handling congestion on the transmission grid resulted. Management of the grid was not supporting all of the commercial transactions and led to serious problems with wholesale trades being cancelled or cut by system operators. New commercially disruptive ways of coping with reliability concerns due to congestion caused serious problems with consummating wholesale transactions. Calculation of available transmission capacity (ATC) by system operators was often inaccurate, leading the system operators the task of cutting scheduled wholesale transactions.

Although contractual sales were made on a point-to-point contract path, the physics of electricity transmission often resulted in unintended movement of electrons into other parts of the interconnected electrical grid. This effect, called loop flow, has the power to disrupt power flows in neighboring control areas; North American Electricity Reliability Council (NERC) authorized system operators to cut transactions threatening the reliability of the system, often resulting in unpredictable trading conditions for wholesale traders. TLR or Transmission Loading Relief became a procedure adopted by operators and approved by the NERC, which resulted in many cut wholesale transactions. The opaque method of arriving at the decision to invoke a TLR or cut transactions added a level of uncertainty and financial risk into the wholesale power-trading environment.

These operational difficulties have led to FERC Rule 2000, which seeks to address these obstacles to developing competitive bulk power markets in the US. It seeks to facilitate non-discriminatory open access of transmission services via Regional Transmission Organizations (RTO). According to the FERC, a patchwork of rules that change from one service territory to another, discrimination by the incumbent utility within their service territory, and high transmission cost as a result of tariffs imposed by each utility are transmission issues which inhibit development of competitive markets.

Order 2000 seeks to resolve these issues through the creation of RTOs. The FERC hopes to improve efficiencies in transmission grid management, remove remaining opportunities for discriminatory transmission practices, improve market performance and facilitate lighter handed regulation. It asks all transmission-owning entities, including non-public utility entities to place control, but not ownership, of their transmission facilities under the control of an RTO. FERC leaves it is up to the transmission entities to develop the RTO in consultation with state officials, allowing industry participants flexibility in the structuring of RTOs. The minimum characteristics are that they show independence, a broad scope and regional configuration, operational authority, and measures to provide for short-term reliability. The FERC also gives some leeway to transmission owner companies in choosing a business model for RTOs. The RTO may be an ISO (Independent

System Operator), whereby operation of a transmission grid is handed over by the transmission owning company to an Independent System Operator or a Transco, an independently owned grid company.

Adding new transmission capacity to the grid has been extremely difficult despite the increased in the number of transactions and planned generation. Difficulties not only with citing, but with a lack of regulatory direction as to how to be compensated for this transmission has led to the new solutions. Combined with new technological developments in micro turbine technology, operational difficulties with transmission has facilitated the development of a sub-industry, distributed generation, whereby micro turbines are generating power at the point of consumption, bypassing the grid all together.

RETAIL DEREGULATION: CUSTOMER CHOICE AND RECOVERY OF STRANDED ASSETS

While a strengthened FERC could provide leadership in the process of deregulating power in terms of its authority in interstate trade, transmission, and wholesale pricing, a major part of the process is under the power of the individual 50 states.

Decisions on how to implement retail access, or customer choice remained the responsibility of the individual states. It is also the responsibility of individual state regulators to decide on how to deal with the thorny issue of stranded costs. Stranded costs can be understood to be uneconomic assets acquired by utilities under a monopoly structure and may include the recovery of the high capital costs of nuclear plants as well as above-market power purchases from QFs, mandated by PURPA. The original projects undertaken by these utilities under a monopoly structure assumed amortized recovery of high capital costs by ratepayers. Similarly, certain contracts, such as above-market power purchase contracts, mandated under PURPA under the regulatory regime, could be understood as stranded costs. The FERC recognized that transition to competitive markets could potentially damage utilities and their ability to provide reliable service during this period and reaffirmed the states responsibility to deal with the issue of compensating utilities for stranded costs. What is the size of stranded assets? Who should shoulder these costs—shareholders, ratepayers or taxpayers, or all three? How should these costs be recovered?

Although each state approached these issues differently, the political will to implement plans for retail deregulation correlates most closely to the level of local electricity costs. At present, four years after FERC 888 and 889, only 17 states have implemented plans for retail deregulation with three others soon to follow. California, Rhode Island and Pennsylvania, states with high power costs, have led the states with their efforts to deregulate the marketplace. Industrial corporations, led by trade advocacy groups like ELCON, as well as political officials concerned by the potential loss of tax base and job loss associated with loss of industry to states with lower cost power, were powerful agitators for change. Rates, which could vary as much as 500% nationally, became a powerful motivating, influence to implement plans for a transition to competitive markets.

The lead in pursuing deregulation on the part of the states was California. In 1994, it issued its Blue Book or Memorandum of Understanding by which it outlined a comprehensive plan for state electricity restructuring. This plan involved a plan for stranded costs, implementation of a market based mechanism and the institution of an ISO, or Independent System Operator. An Independent System Operator (ISO) runs the exchange's transmission, which is a separate entity from the exchange. After many debates, the State of California adopted a PoolCo model method of establishing prices. The PoolCo, or Calpx, became not only a mechanism for buying and selling power, but a mechanism for recovering of stranded costs by the three large IOUs in the state. The three large IOUs in the state: San Diego Gas and Electric, Socal Edison and Pacific Gas and Electric were mandated to sell and buy power exclusively from the California Exchange (CalPx). In 1998, customers were given the choice of choosing suppliers with their rates capped for 4 years at 10% of the average of pre-restructuring costs for their rate class. Legislation (Assembly Bill 1890) allowed for the financing of this rate reduction by allowing the utilities to issue $10 billion of bonds, collateralized by Customer Transition Charges (CTCs). CTCs were recovered from the Calpx by the taking the difference between the customer charged rate and the spot rate earned on the Calpx. The higher the Calpx discount to the customer rate, the greater amount of CTCs earned by the utilities. The utilities have a four year time limit to earn these CTCs through the Calpx mechanism, when they would also no longer be required to sell into the Calpx. Southern California Edison and Pacific Gas & Electric also divested of 13 generation plants in line with the Public Utility Commission's divestiture policy.

Other generators, including out-of-state generators, were eligible to buy and sell power from the Exchange, although it was not mandated, since the state had no jurisdiction over them. Independent power producers (IPPs), municipal utilities, utilities located outside of California, aggregators, etc., have the option of buying from or selling electricity through CalPX or selling directly to a customer without going through CalPX.

The northeastern part of the U.S. is progressing in developing retail competition, led by Pennsylvania, New Hampshire, Rhode Island, New Jersey, and New York. Each public utility commission has created a different plan to offset stranded costs, often ordering divestiture of generation assets. Like California, many plans involve the "securitization" of stranded assets, whereby utilities issue bonds to recover stranded costs up front, with fixed discounts to a prior average calculated as the cap of customer prices, i.e. shopping credits.

The transition on the state level has been very slow. States who have implemented retail programs have often instituted complex, re-regulated solutions, which in reality delay the introduction of real competition.

FEDERAL ELECTRICITY REGULATION

Several outstanding issues in preventing the development of competitive markets must be addressed by legislation. Several bills have now been introduced into Congress to address some of the jurisdictional issues, which hamper trade development. FERC 888 and 889 did not completely open the transmission system. There are certain barriers to wheeling in regions with Federal electric utilities such as Tennessee Valley Authority (TVA), Bonneville Power Administration (BPA) and other Power Marketing Agencies. Since FERC 888 and 889 transmission rulings applied only to Investor owned utilities, additional legislation is needed to open up transmission systems of these other entities. New legislation proposes enforcement of reliability rules and allowing the federal government to enforce reciprocity rules. For example, if a state, which has not opened up its market to deregulation, wish to sell power in a neighboring state in which the market is open, proposed legislation address enforcing reciprocity. Other provisions include the repeal of PUHCA as well as parts of PURPA, such as required power purchases. Passage of federal legislation can do much to open the landscape to competition.

MARKET DEVELOPMENT AND THE EMERGENCE OF REGIONAL HUBS

The fractured nature of regulatory jurisdiction has greatly complicated the process of deregulation in the United States. Initial expectations from some that the power industry deregulation process might follow that of natural gas in terms of regulatory change and market development did not materialize for this reason. In natural gas, the well-established interstate pipeline system gave federal authorities primary responsibility to the FERC to drive the deregulatory process. A competitive spot market materialized, which grew rapidly, along with, an over-the-counter (OTC) swap market and a deep, liquid futures and options on futures market, which now extends out to three years. Regarding development in power, differences between states have been dramatic in terms of initiatives to implement retail deregulation, bringing the end-user into the competitive market place. Additionally, the real time, non-storable physical nature of power requires an operational transmission and reliability infrastructure for the development of fungible commodity instruments to develop.

The initial market activity was an offshoot of systems coordination between neighboring utilities, whereby excess generation in one utility was sold to a neighboring utility if the cost was cheaper. FERC regulated these transactions, mandating that the savings be split 50-50. Other arrangements included sales of power from IOUs or independent power producers to municipal utilities and rural cooperatives, also regulated by the FERC. EPAct and then FERC 888 and 889 encouraged the formation of unregulated bulk Power Marketing entities and wholesale power-marketing transactions began to increase.

Trading in regional centers gradually began to develop, each idiosyncratic and a function of its own physical infrastructure. The wholesale marketplace developed the fastest in the Western U.S. In addition to the progressive stance on deregulation taking by California, several other factors encouraged an active wholesale market in the West.

The Western System Power Pool (WSPP) is an organization formed in 1987 to encourage additional efficiency in the interconnected power operations of member systems in this area. It did so primarily through enhancing members' market knowledge, establishing methods of flexible pricing, and through the standardization of contract terms and conditions for wholesale trading. Standardization was a key factor in promoting wholesale market development. In contrast to other regions, where utilities signed individual contracts with counterparties, members of the Western System Power (WSPP) operated under a standard contract, eliminating the need to open up individual contracts with other traders. The FERC approved a master WSPP agreement in 1991, which covers sales of capacity, energy and transmission by WSPP members. It also provided information to its members as well as an electronic bulletin board. In 1995, it contracted with Dow Jones to build the first proprietary electronic trading platform for electric power.

Trading in the West also developed quickly due to the different patterns of power consumption, i.e. winter-peaking characteristics in the Northwest U.S. and summer-peaking in the South led to an active North South trade, supported by a powerful high voltage transmission infrastructure. Huge hydroelectric dams in the Northwest produced plentiful power from spring runoff, exporting it south during peak summer consumption. During the winter, when hydropower was less plentiful, power was imported from the South. An active trade in day, forward, weekly, balance of week, monthly, balances of month, quarterly and yearly strips began in this area.

The need for price transparency in this evolving marketplace led to the creation of price indexes as a source of independent price information. Dow Jones created a series of price indexes for these emerging hubs. The first one was for COB, the California Oregon Border, and a developing trading hub in the Northwest. In the West, Palo Verde, and Mid-C followed this one. These indexes were the first of their kind and were constructed by entering into contractual agreements with key wholesale traders in a region, recording all transactions and prices, weighting the averages and publishing a daily index for firm and non firm next day power. These indexes were fully auditable and became the first form of independent price discovery for the area. Other indexes were constructed for additional developing hubs: Palo Verde in Arizona, Mid-C for the trading hub developing from trading hydroelectric power from the five non-federal hydroelectric plants along the Columbia River.

Gradually, an active wholesale market in the West included an active daily pre-scheduled spot market, weekly market, balance of week, monthly, balance of month, quarterly, and yearly strip market. A standard size wholesale transaction for on-peak trades (16 hours daily, 6 days a week) is generally 25 MWh.

Having previously announced its intention to enter the power market, the increasing activity in the West led the New York Mercantile Exchange to introduce its first electricity contracts in March of 1996: the COB and Palo Verde Contracts. These contracts were based on delivery at the California Oregon Border (COB) and Palo Verde, Arizona. These contracts were designed to be used as instruments for price hedging, or to mitigate financial risk associated with price volatility.

An organized futures exchange regulated by the Commodity Futures Trading Commission, the New York Mercantile Exchange is a membership organization with a clearinghouse that assumes counterparty credit risk for the performance on contracts. Trades are collateralized with good faith margin deposits of cash or government securities and are marked to market daily through the 57 clearing members who collect margin from and guarantee the trades of the traders and customers they represent. These clearing members include some of the most prominent and widely respected names in the banking and financial services industry. Financial and physical performance are guaranteed on trades are guaranteed by the Clearinghouse. The relatively small size of the contract allows for the participation of smaller buyers and sellers.

Although the contracts have been modified somewhat in size and contract unit since the original launch, the contracts are based on a monthly delivery unit of 432 MWhs of firm electricity to be delivered as follows: one MW per hour delivered over a 16 hour peak period, 6 days a week (6X16) during the specified delivery month. Eighteen delivery months trade simultaneously for each contract. Contracts may be offset by taking the opposite position prior to the expiration or the contract or may be settled by taking or receiving physical delivery. Power for the COB contract is deliverable into the California Border; power for Palo Verde is deliverable into the Palo Verde Switchyard in Arizona. American style options on both COB and Palo Verde futures contracts are also traded.

In the Midwest, the Cinergy control area became an active hub for wholesale trade. An investor owned utility (IOU), Cinergy's control area covers 25,000 square miles and 7,000 miles of transmission lines. Its central location, having first tier interconnects with 11 other control areas and 2nd tier interconnects with 25 control areas made it natural hub for wheeling power throughout the Midwest. An active trade developed in and "Into Cinergy" product, which meant that the seller had a choice to deliver power anywhere within the Cinergy control territory. The trade conventions in this part of the country were slightly different from that of the West. Larger blocks of power were standard units of trade. A standard wholesale on peak transaction generally was for 50 MWh delivered over 16 hours a day, 5 days a week.

The Entergy control area also became another dominant market hub, encompassing an 112,000 square mile area in Arkansas, Louisiana, Mississippi and Texas with 15,500 miles of transmission lines. It is interconnected with 11 utility systems including TVA, Southern Co. and Central & Southwest. An active wholesale market in the "Into Entergy" market has developed into a liquid wholesale power market, using similar conventions as the Into CINergy product.

Consequently, the New York Mercantile Exchange launched two additional futures contracts in July of 1998: Cinergy and Entergy. Due to different trading conventions in the Eastern U.S., the contract size for both contracts was for 732 Mwh deliverable over a month. This equates to a delivery of 2 MWh, 16 hours a day, 5 days a week over the course of the specified delivery. The Cinergy contract called for power to be delivered at the seller's choice of location within the Cinergy control area. Similarly, the Entergy contract called for seller's choice of delivery within the Entergy control area.

The physical infrastructure of the northeastern U.S. markets, influenced by its geography and population density, was designed to deliver power relatively short distances in tight interconnected areas. Larger, high voltage transmission lines were built later to connect nuclear facilities to distribution areas as well as lines to transport Canadian hydropower south. Tight power pools developed in PJM (Pennsylvania, Jersey, Maryland), New York and New England. Transmission congestion associated with an active wholesale market became a key issue to be contended with in this region. The PJM (Pennsylvania Jersey Maryland) became a hub and a robust wholesale market developed on a "sellers choice" model, which allowed the seller to deliver anywhere on the 350 KV line within PJM. When congestion pricing was instituted in the region in January of 1998, the area had a sharp decrease in wholesale power transactions, since pricing became broken down to 2,000 buses in the location. When prices were aggregated in three buses, wholesale market returned somewhat. The New York Mercantile Exchange launched a contract for this location as well, structured similarly to the Cinergy and Entergy contract in terms of contract size and structure, but for delivery on to the Western hub of the PJM Interconnection.

RISK MANAGEMENT: DEVELOPMENT OF MARKET TOOLS

The development of markets includes trading in hourly markets, daily markets: forward markets are traded weekly, balance of week, monthly, balance of month, quarterly, and yearly. The percentage of deals that are purely financial with no physical settlement are still relatively small and include the futures markets. However, the rapid divestiture of generation assets as well as over 100,000 MW of new planned generation is creating a new class of merchant plants with risk management needs relating to the managing of operating margins or the spread between fuel costs and power sales. Many power developers are subcontracting the merchant function to skilled trading organizations, which will manage the physical and financial purchase of fuel and sale of power on behalf of developers. This management involves using futures, forwards and other structured financial products to lock in the margin between fuels and power output. Transmission and distribution companies need to manage their forward purchases of power. As transmission and operational issues become worked out on a regulatory level, wholesale markets should become deeper and more liquid. Also broadening the regions traded by means of RTOs should broaden the limits of fungibility also facilitating the development of forward financial markets.

Given the level of wholesale market development in the U.S., the design of futures contracts for electricity has posed special challenges. The design of futures instruments is based on standardized terms and conditions for a particular grade of commodity, deliverable at a certain location over a certain time period. The standardization of terms concentrates buyers and sellers, concentrating liquidity and narrowing bid/offer spreads. Non-standard grades are priced for deviations in grade and location from the standard contract specifications. A basis market develops as price discovery extends to other grades. Natural gas is an excellent example of this. While the delivery point for the New York Mercantile Exchange natural gas contract is the Henry Hub in Louisiana, this location is used as a reference point from which to price other deliveries of natural gas in other locations. The differentials related to the price fundamentals of transportation.

Electricity deviates from this model for several reasons: because electricity is largely unstorable and is produced and consumed at the same time, it is sold in many different discrete time units. Electricity is sold in half hour increments in Australia and UK, one hour minimum increments in the U.S., on a next day basis, weekly, balance of week, monthly, balance of month, quarterly, yearly and so on. Because of the real time nature of electricity, the market is granular by its nature. And because of the physical nature of power, transportation involves heat loss. Therefore, the transportable distance of power is limited, making basis calculation beyond a certain distance meaningless. Compounding this granularity are the current operational difficulties with transporting power, which FERC is addressing in its latest rule. All of these factors require the creation of multiple contracts. Additionally, the futures contracts are financial instruments designed to hedge term power transactions. Price volatility due to generation shortages as well as chaos in the transmission markets has discouraged long term trade, which the futures contract, is designed to hedge.

Factors promoting increases in futures should be expected to be entrance of more players into the term market place from retail end users as retail markets develop, the creation of large scale RTOs expand the fungibility of traded electricity products.

Other futures contracts launched by other exchanges include a TVA contract and a COMED contract by the Chicago Board of Trade and the Twin Cities Electricity contract by the Minneapolis Grain Exchange.

EMERGENCE OF ELECTRONIC TRADING PLATFORMS

Due to many of the reasons enumerated above, much of the forward electricity transactions are done in the OTC market, where many customized trading products exist. These products can vary widely from region to region, depending on physical market trading conventions and transmission infrastructure. The number of products and regions traded and the need for firmness in executing transactions has made trading on electronic platforms increasingly attractive in the power industry. Firms have indicated a desire to automate their entire business from aggregating retail transactions to maintaining portfolios, to calculating risk exposure to the execution of orders to hedge this risk. Electronic trading platforms facilitate this. The extraordinary developments in Information Technology and the growth of the Internet have led to the development of numerous transaction platforms for cash OTC cash trading and swaps as well as futures instruments.

The platforms fall into a number of categories. Order Book Matching systems are systems run by third party companies, providing a platform for matching of customer limit orders for differing size, tenor and location. There are now five such operational platforms. All of these systems allow for bilateral trade and provide a kind of credit filter, whereby participants are able to pre-choose their counterparties. The company providing transactional service does not guarantee performance on transactions.

A second model for transactional platforms currently in operation is a customer management system, whereby a company becomes a counterpart to its customer's transactions. There are two such platforms in active use now in the power business. While it may be possible for customers to execute transactions with the operating company, there is no independent trade monitoring function. Non-regulated exchanges represent another class of trading platform, i.e. the Calpx and APX are examples of this kind of platform, providing unregulated spot physical and forward transactional capability.

Regulated futures exchanges, such as the New York Mercantile Exchange form another distinct class of electronic transactional system. The NYMEX ACCESS platform, originally developed as an after-hours electronic trading platform for regulated futures and options on futures contracts, now has day-time trading functionality. It is also the exclusive trading platform for the New York Mercantile Exchange electricity futures and options contracts. The Exchange has announced the formation of eNYMEX, an Internet-based, physical commodity OTC trading and clearing platform, which incorporates order routing to the futures markets, allowing for net portfolio marginning. It will allow for efficient side-by-side hedging, using both futures and options on futures contracts as well as OTC instruments.

The number of new platforms being developed to trade energy products is now increasing dramatically. While the number threaten to fragment the market, they also provide new promise for efficient and cost-effective solutions to trading physical commodities as well as the execution of financial risk management products. Brokers of OTC products are now transitioning customers to use affiliated platforms and have assumed a new role: monitoring different platforms for the narrowest bids and offers. As protocols become standardized and these platforms become Internet-based, there are opportunities to automate this function to find the best combination of products at the best prices, integrating the best of all including clearinghouse protections of established exchanges. Liberalization of regulatory procedures is freeing organized regulated exchanges to create products with less lead-time and to create new classes of instruments as well.

ELECTRICITY DEREGULATION IN EUROPE

In 1996, the European Union issued an European Electricity directive, organized to deregulate the power sector in Europe. The goal became the creation of a single energy market based on open and competitive markets, with the objective of lowering costs due to prior isolation of national markets and lack of intra community trade. Addition goals were security of supplies, diversification and a closer integration of internal markets.

Essentially the directive mandated a competitive market for generation whereby in February of 1999, any producer can build new electric power generation and laid out the procedures for approvals for new projects. It also mandated that depending on customer size, choice would be available to 26% of the retail market in 1999, 28% in 2000 and 33% in 2003. The requirements of this phased in process would be in three steps. The first step is calculated as customers who have a total annual consumption exceeding 40 GWh, calculated to be 26% of the market. The second step, the requirement is to open the market for customers with a consumption level of 20 GWh, bringing the competitive market to 28% of the consumers. The third step is to extend choice to customers whose consumption is over 9 GWh, which in total amounts to a market opening of 33%. The Directive also gives guidelines for public service obligations, outlines procedures guiding reciprocity. Like FERC 888, it requires the functional unbundling of generation, transmission and distribution.

The directive lays out a plan for three methods of transmission access: regulated third party access, negotiated third party access or the single buyer model. So far, all member states have opted for regulated or negotiated third party access. It also gives guidelines for dealing with stranded costs, including social obligations as well as fuel and power purchase agreements. Cross-border transmission pricing is regulation at European Union level.

Germany, whose electricity prices were the highest in Europe, has attracted the most attention with its decision to open its markets to 100% of its retail customers in February 1999 with no provisions for compensation for stranded costs. The German electricity markets, characterized by over capacity in both generation and transmission have dropped in price by 60% in a year. Although some large American marketing companies active there have complained about difficulties with non-discriminatory access to the transmission grid, they have also enjoyed success in capturing industrial customers.

Germany has also taken a leadership role in Continental Europe in the creation of a futures exchange. Four cities were interested in becoming the site of a new futures exchange. Stuttgart, Leipzig, Frankfurt and Hannover all pursued the goal of becoming the city of a new futures exchange. The German government ultimately approved Frankfurt as the location of the Exchange, which would be an electronic exchange created by collaboration between Deutsche Borse and Eurex. However Leipzig decided to proceed with its own exchange, and the Frankfurt-based European Energy Exchange will now compete against the Leipzig Power Exchange. Ahead of the launch of both exchanges in summer 2000 there were talks about a merger. Most expect a merger to take place, as the market consensus is that there is insufficient liquidity in the German market to support two exchanges.

France in contrast, has a large nationally owned company, Electricite de France (EdF), whose generation is 95% nuclear. It has taken a position of stonewalling domestically on opening markets, while actively exporting across the continent to Netherlands, UK and Germany. EdF has also purchased bought London Electricity (the UK REC) and has a joint trading venture with Louis Dreyfus in London. Although France is introducing new legislation to implement the changes required by the EU Directive but there is little prospect of any significant competition in generation or supply.

CONCLUSION

The evolution of more efficient power markets globally has led to the pursuit of free market policies worldwide. An evolution of power markets from vertically integrated monopoly structures to ones marked by the unbundled functions of generation, distribution and transmission is in the process of taking place with associated development of competitive market structures. In some cases, this is a phased-in process marked by the peeling away of layers of legacy regulation, like the US. In other cases, regulatory directives have stimulated radical, dramatic change as in Germany.

Different market structures have evolved from various types of pools to active OTC bilateral trade and organized futures exchanges, providing instruments to trade physical and financial power. New technology has accelerated the creation of efficient markets by new generation technologies. Astounding advances in Information technology have also transformed the speed of handling transactions of energy trades. The changes set in motion are powerful and inexorable. The next several years should witness a period of integration and implementation of these new markets.

CHAPTER 3

Weather Trading: Raising the Temperature in Monte Carlo

By Andrew Feachem

INTRODUCTION

Weather derivatives are the first financial tool available to risk managers to stabilize earnings volatility caused by the unpredictability of the weather. These financially settled products are structured as temperature swaps and options and are used to hedge volume-related risks caused by extreme weather conditions. While traditional weather damage insurance has been available to businesses for years, protection in the form of a derivative designed specifically for non-catastrophic weather conditions has only recently become available. When used together with traditional price hedges, properly structured weather derivatives can reduce cash flow volatility, lower financing costs and stabilize revenues.

The market started during the summer of 1997 when two US power companies, realizing that they had opposite weather exposure, entered into the first weather derivative swap contract for the upcoming winter season. Since then, the market in the US has grown rapidly to an estimated 2,500 deals with a value at risk (VaR) of around $5 billion. Markets are also developing in Europe, East Asia and Australia with predictions of the European market being worth $8 billion within the next two years. This chapter provides a thorough background to these innovative products and some perspective and scope on the future of weather derivatives.

A UNIQUE RISK HEDGE

A weather derivative's most distinctive property is its ability to address volume risk, as opposed to other traditional risk management tools that address price risk. For example, an electric utility can utilize a weather derivative that generates payments when extreme summer weather requires ramping up of high cost generators. Conversely, this same electric utility can benefit from a different weather derivative that generates payments when mild summer conditions result in decreased load.

The following diagrams briefly demonstrate the stabilizing effect a weather derivative can have on a utility company's revenues:

FIGURE 3.1

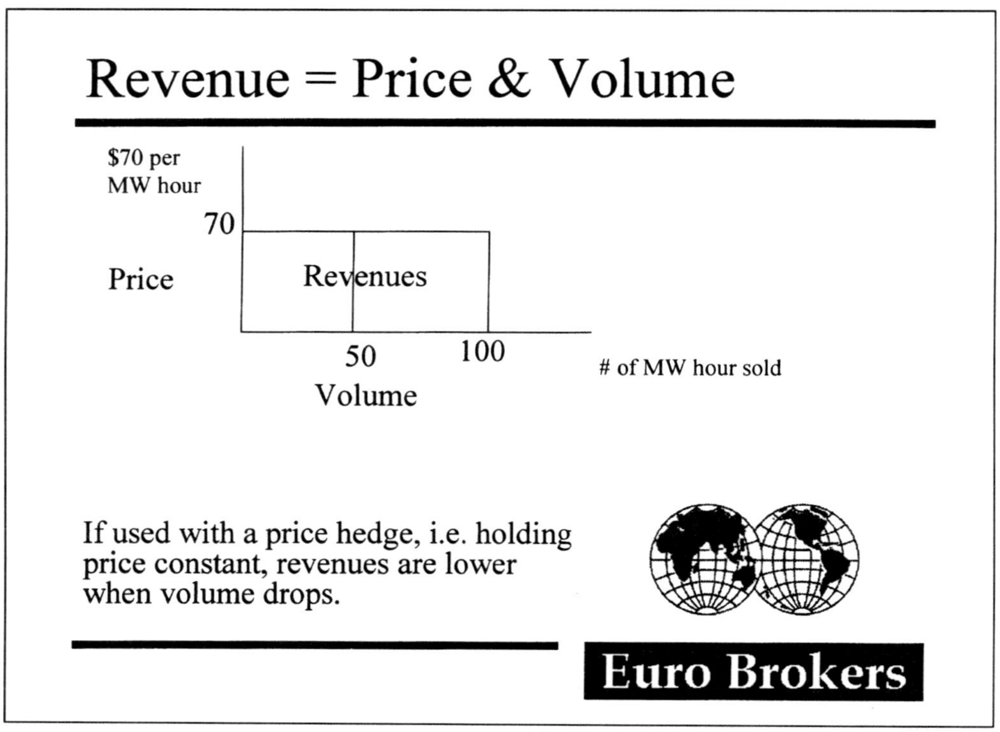

Revenues are a function of price and volume. Using the above example, on the assumption that the utility company has an electricity price hedge in place that has the effect of holding power prices constant, if the volume of electricity sold drops from 100MW/hr to 50MW/hr, due to low demand caused by a warm winter, the end result will be a halving in revenues.

If we combine the traditional price hedge with a winter Heating Degree Day put option, the overall effect on revenues can be seen below:

FIGURE 3.2

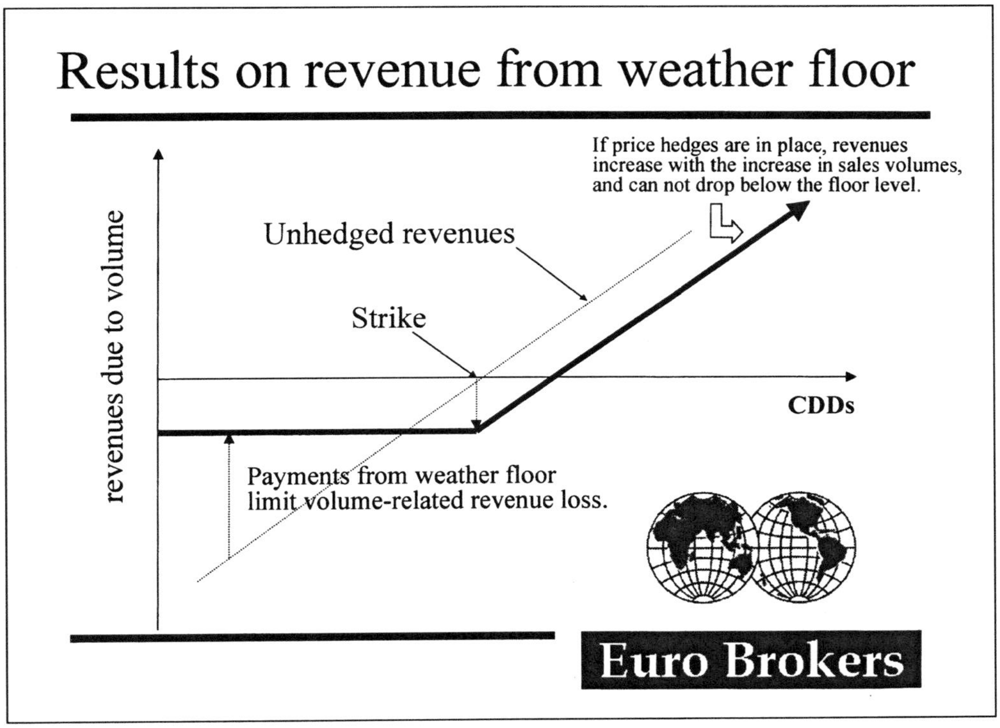

The option will payout to the utility company the warmer it becomes over the winter season, i.e. as the heating degree days fall. The result is a limit, or floor, on revenue loss whilst maintaining the upside potential of a cold winter.

Price and volume affect revenue and profitability. A hedge program that combines weather and traditional price hedges enables risk managers to address both components that affect the bottom line. This will result in more efficient cash flow management leading to an overall improved corporate financial position.

SPECIFICS OF WEATHER OPTIONS

A key feature of the weather derivative market is that there is no underlying instrument as, obviously, there is no physical market in the weather (you can't buy rain in London for January, although you can count on its occurrence). Derivative products are used to hedge exposures built up in its underlying. As there is no such underlying in this market, weather derivatives are used to hedge against the risks that are affected by the weather, for example, the risk that energy demand decreases.

With the absence of an underlying instrument traded, the market's best estimate of the weather will be its swap level, or futures price if such a contract exists on an exchange. The swap and futures level will change as the market's estimate of the weather changes, taking into account the information contained within short and long-term meteorological forecasts. The details of how swap and other derivative contracts pay out will be outlined later in this chapter.

Unlike many commonly traded derivative instruments, weather options will settle on an accumulated index, such as heating degree days (HDDs) and cooling degree days (CDDs). By definition then, these options will always be European in style and not be exercised until the end of the contract period.

INSURANCE VERSUS DERIVATIVE

Often people describe those who enter into weather derivative contracts as "gambling on the weather". Although there is sometimes an element of this, the major use of a weather derivative is to allow a company to gain some protection against adverse weather conditions. They are therefore commonly referred to as insurance products and share some similarities to insurance policies, with there also being some striking differences.

Weather derivatives, like insurance policies, almost always have a maximum payout—not only because most participants, particularly from the insurance/reinsurance industry, require knowledge of their maximum risk exposure, but also as the weather is finite. We know that temperatures could not exceed 50(C for an entire month in London, so we know that there will be a natural boundary to the value of which a cumulative temperature index will reach.

The two principal benefits of a weather derivative over traditional insurance are:

- You do not need to prove a financial loss to receive a payment from a weather derivative. An insurance policy will not only require a specific weather event to occur, but also to demonstrate that a financial loss was incurred as a result of this weather event before the insurance company pays out. A weather derivative will pay out simply if the weather event occurs. For example, a Florida citrus farmer who claims against an insurance policy because frost has damaged his crop will have to wait until the end of the growing season to demonstrate that his crop is ruined before he is compensated. If he instead chooses protection with a weather derivative and there is a frost, not only will he be compensated by the derivative, he can save a portion of his crop by harvesting early before it is entirely ruined and sell it for concentrate.

- Further, derivative contracts are generally more efficiently priced than insurance policies, with weather derivatives often trading at, or near, their fair value (i.e. where net expected returns to the counterparts is zero).

Finally, insurance products are better suited to protecting against extreme one-off weather events, catastrophes such as hurricanes, whereas weather derivatives are most often used to protect against milder movements in weather conditions over a season.

COMPONENTS OF A SUCCESSFUL MARKET

As in all markets, there are several conditions that need to be met in order to promote the chances of liquidity. Although the weather market is illiquid when compared to, for example, interest rate derivatives, it does satisfy many of these conditions with the rest to follow in the near future.

- **Large number of market participants:** The potential for this market was first realized by the energy sector just a couple of years ago, so it is not surprising to find that the current most active participants in weather derivative trading are power marketers/utilities. US companies such as Koch, Enron, Aquila and Southern have been involved, on at least one side, of the vast majority of structures that have traded to date. Most end-user deals have also been within the energy industry. The significance of this sector will remain in the future, but we are now seeing a greater number of end-user deals taking place in other industries. For example, two transactions were recently announced in Japan; one for a ski resort protecting against a lack of snowfall and the other for a supermarket protecting against a fall in shoppers caused by bad weather.

There has also been rapid growth in financial institutions participating, most noticeably within the reinsurance industry. Reinsurance companies are in the business of taking on risk. They are able to do this by building a very large portfolio of diversified risks and since weather patterns are unlikely to be correlated with the traditional risks that they have on their books, taking on positions in weather derivatives is a natural for their business. They are able to take very large exposures and will lend capacity to the market that the current energy participants cannot.

There are a fairly wide variety of players in the market now and it is diversifying rapidly. European and East Asian participation will grow significantly in 2000 and we will see more deals transacting in the non-energy sectors. These are positive signs for future liquidity.

- **Reliable data for indices:** Reliable data is not only important for establishing indices that participants will trust and actively trade derivative contracts from in any market, but especially in the weather market where historical data is essential for the pricing of structures. In the US, highly reliable and unbiased data is published by the National Weather Service (NWS). It is generally possible to obtain at least 50 years of historical data for over 300 primary automated sites and 2000 secondary "co-operative" sites nationwide at extremely reasonable cost. All this helps to provide transparency to all those wishing to enter the market, increasing confidence, the result of which is elevated participation.

  We have seen the result that opaque data sources have had on the development of European weather derivatives. There is no doubt that the quality of historical and current data from the respective European meteorological services is of a high standard. Sadly it is also of very high cost. Originally the UK's Meteorological Office had intended to sell 50 years of data for eight UK locations for around $30,000. This had the immediate effect of preventing all but the largest companies from getting involved in weather derivatives—if you are an end-user looking to put in place a one season protection, $30,000 to price the structure is a little unreasonable. This cost has now fallen to $6,000. A vast improvement, which has helped to boost the market in the UK, but still a far cry from the $1,000 one would pay for thousands of sites across the US.

Data transparency is improving and in some cases the data quality and cost are excellent, as in Japan and the US. Competition in data provision is developing in Europe and hopefully we will see a significant reduction in data costs over the year.

- **No market manipulation:** There will be little confidence, leading to diminished liquidity, if there is the feeling that the market can be manipulated. With weather derivatives, this is virtually impossible. No one can corner the market for CDDs in Philadelphia for the summer, or try to squeeze HDDs in Boston in December. The weather cannot be controlled, stored, or transferred from place to place. Mother nature provides a level playing field for everyone.

- **Attractive to both natural hedgers and risk takers:** Leading on from the first point, it is important for liquidity that we see increasing participation from natural hedgers, or end-users, in order to provide a basis for a secondary, or re-traded market. Potential natural hedgers are in abundant supply with approximately 20% of a country's GDP being at the mercy of the weather. Fortunately there is an ever-growing contingent of risk takers, helped by the fact that volatility, driven by frequently released weather forecasts, exists in the market. Volatility is essential as it increases the value of options and swaps to risk takers in any market. As mentioned above, reinsurance companies, investment banks and some power companies already participate in the market as risk takers and in the near future we will see hedge funds taking advantage of the uncorrelated risks that weather derivatives offer.

TYPES OF INDICES FOR CONTRACTS

Generally, as long as there is a reliable source of data, it is possible to base a weather derivative contract on any kind of weather activity. Listed below are a sample of the most commonly traded and talked about indices on which weather derivatives are based on:

- Heating degree days (HDDs).
- Cooling degree days (CDDs).
- Absolute temperature, such as daily maximum temperature (T_{max}) and daily minimum temperature (T_{min}).
- Precipitation, such as actual rainfall, snow pack and snow fall.
- Dual commodity structures, for example a weather contract which is linked to a gas, or electricity contract.
- Wind speed, which could be used for offshore storm protection.
- Distance of a hurricane from a specific location.
- Misery, a combination of temperature and humidity.
- Growing degree days.

HDDS AND CDDS— THE MOST ACTIVELY TRADED PRODUCT

Most weather derivatives traded are swaps and slightly out of the money options that settle against cumulative degree day indices. These products are customized to cover individual cities, or baskets of cities, with risk periods ranging from a few months to several years. To date, these types of contract have proved to be very popular with the US energy industry. The vast majority of deals traded since the market's inception have been of this kind, with most providing seasonal winter or summer coverage.

Degree days are the number of degrees the midnight to midnight daily average temperature deviates from a base of 65 degrees Fahrenheit. CDDs reflect the "cooling" requirements during summer months and HDDs reflect the "heating" requirement during winter months.

The degree days for a single day can not be a negative number. Therefore, heating degree days are calculated as the greater of zero, or 65 minus the midnight to midnight average of the high and low temperature for the day. Cooling degree days are calculated as the greater of zero, or the midnight to midnight average of the high and low temperature for the day minus 65.

For example, if the average temperature for a summer day is 85 degrees, then the number of CDDs for this day would be 20. If the average temperature for a summer day is 62 degrees, then the number of CDDs for this day would be 0.

Degree days contracts are settled against the cumulative degree days over the contract period. To calculate the value of the degree days index, you would measure the degree days for each day within the contract period and sum them up. By definition then, you will have to wait until the end of the contract period to know the full value of the cumulative degree days index on which to settle (unlike options in many other markets where it is possible to exercise them before the expiration date).

The rationale behind using a base temperature of 65 F in the degree days calculation is that a homeowner will turn on their air conditioning above this temperature, hence cooling degree days, and turn their heating on below it, hence heating degree days. It is unreasonable to assume that 65 is the magic number in all locations across the US, let alone across the globe. It is, however, convenient that all participants have agreed to use 65 F for the US in order to standardize as much of the contract as possible. Outside the US, it is widely accepted that 18 degrees Celsius will be used as the base temperature for the degree days contracts.

Growing degree days (GDDs) lend themselves naturally to the agricultural industry. They function in a similar way to HDDs and CDDs and are used to estimate the growth and development of plants and insects during the growing season. The basic concept is that development will only occur if the temperature exceeds some minimum developmental threshold, or base temperature (Tbase). The base temperatures are determined experimentally and are different for each organism (see tables below).

TABLE 3.3 Temperatures at which crops begin to grow

Tbase	CROP
40 F	Wheat, Barley, Rye, Flaxseed, Lettuce, Asparagus
45 F	Sunflower, Potato
50 F	Sweet Corn, Corn, Sorghum, Rice, Soybeans, Tomato

TABLE 3.4 Temperatures at which insects develop

Annual Reduction Rates	Number of NO_x Facilities	Number of SO_2 Facilities
2%	93	10
2 to <6%	70	16
6 to <10%	89	3
10 to <16%	118	11
Total	370	40

GDDs can also not be negative and are calculated as the greater of zero, or the average of the midnight to midnight temperature minus Tbase. For example, if we wish to calculate the GDDs for wheat on a day where the average temperature is 65 F, the GDDs will equal 25 (65-T_{base} = 65-40 = 25). As before, a weather derivative structure would settle against the cumulative GDDs over the contract period.

Often when temperatures rise to high values, the benefits on the growth rate of organism will be diminished and sometimes reversed. Modified growing degree days will take into consideration this effect. If the daily maximum temperature is above 86 F, it is reset to 86 F. If the daily minimum is below 50 F, it is reset to 50 F. Once the maximum and minimum temperatures have been modified if needed, the average for the day is computed and compared with the base temperature (usually 50F). Modified growing degree days are typically used to monitor the development of corn, the assumption being that development is limited once the temperature exceeds 86F.

WEATHER DERIVATIVE STRUCTURES

Weather derivatives are very flexible and can be tailor made to hedge very specific risks. Consequently there are numerous types of contract that are commonly talked about, and a number that are actively dealt.

Commonly traded structures:

- **Caps.** A call option on a weather index where the buyer of the option receives a payment if the defined weather index exceeds the strike level at the end of the contract period. The amount of the pay out is generally related to by how much the index exceeds the strike level. The buyer pays the seller a one of premium for this contract.

- **Floors.** A put option on a weather index where the buyer of the option receives a payment if the defined weather index falls below the strike level at the end of the contract period. The buyer pays the seller a one of premium for this contract.

  The diagram on the next page illustrates how a HDD floor would pay out to the buyer of the contract. In this case, the buyer is looking to protect against a drop in earnings from a mild winter, while allowing for the increased earnings from favorable weather conditions. Notice how the pay out increase the further below the strike the index settles. The floor buyer receives payments from the floor seller if the HDDs are below the strike level.

FIGURE 3.5

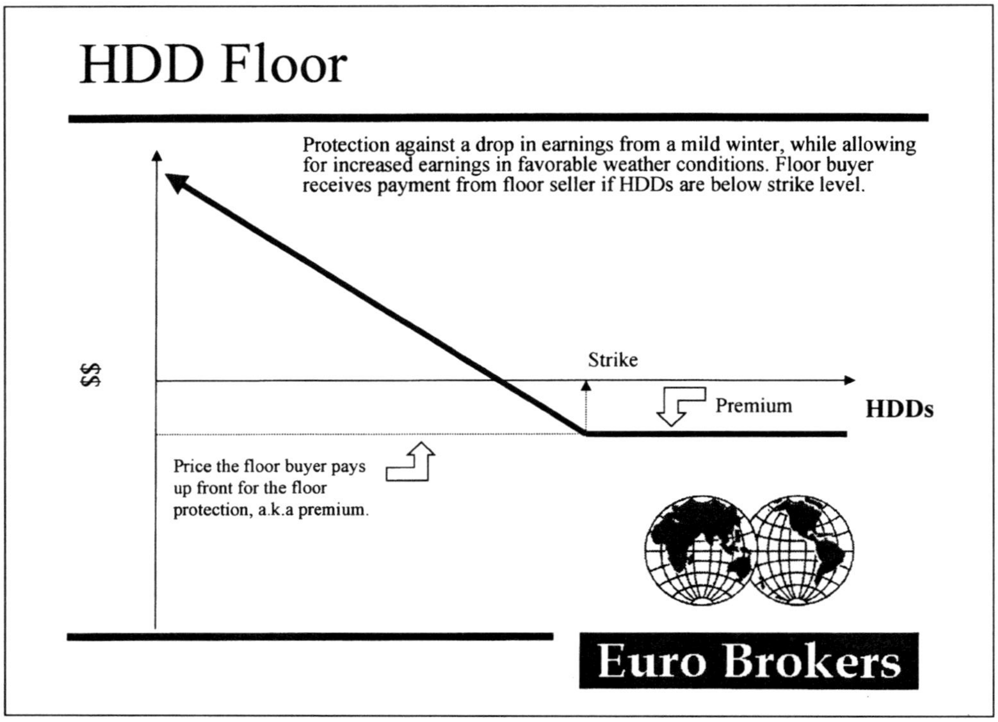

- **Swaps.** A financial contract which has the effect of "swapping" cashflows between the counterparties, depending on which side of the agreed swap level the defined weather index settles at the end of the contract period. The buyer of the swap receives payments from the seller if the index settles above the swap level. Conversely, the buyer makes payments to the seller if the index settles below the swap level. There is no exchange of premium for this contract as both the buyer and seller have the potential to make or lose money equally on the swap contract.

  In the diagram on the next page, we see the effect on a utility's revenues when they sell a CDD swap over the summer. In extreme summers, the utility sells more electricity, increasing volume related revenues. To protect against a mild summer, the utility can sell a swap, giving away some upside profit for downside protection. The result is earnings stability.

FIGURE 3.6

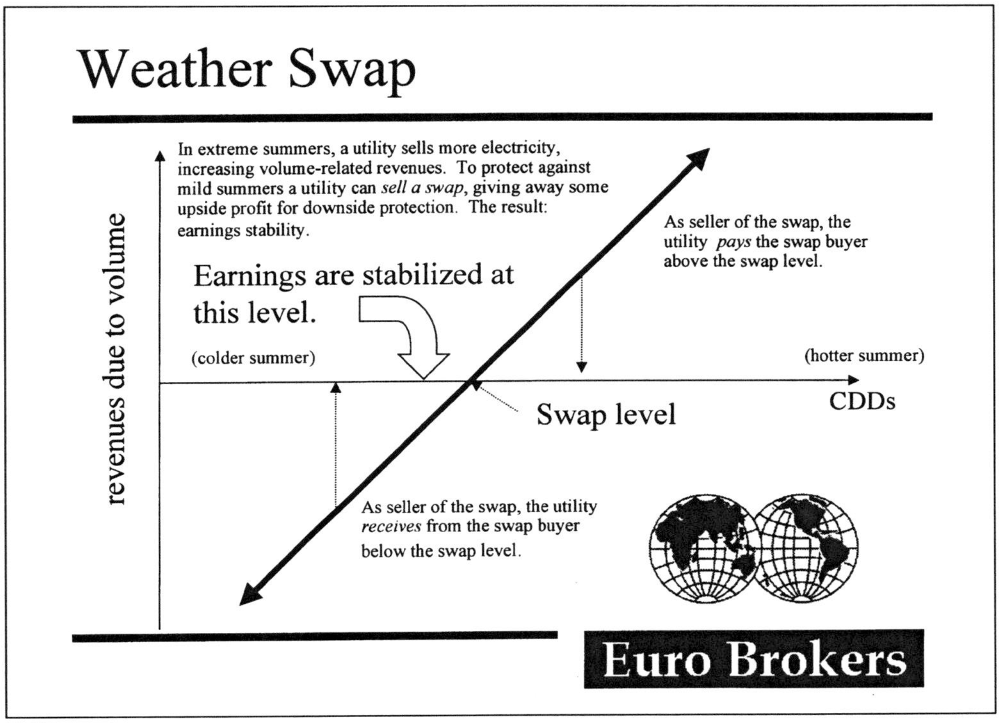

- **Collars—Cost-less, or Low Cost.** Similar to the other options, a collar allows a business to protect against adverse weather conditions at little or no cost by giving away some of his upside potential. The buyer of collar essentially buys a call option, cap, and funds this by selling a put option, floor. Naturally, the strike of the call will be greater than the strike of the put, and if the collar is cost-less, then the average of the strikes should be equivalent to the swap level. The following diagram depicts the effect on revenues of an energy company entering in to such an agreement. Over the summer, he wishes to protect against mild temperatures by buying a CDD floor. He will fund this option by selling a CDD cap, which will have a strike higher than that of the floor, such that the value of the cap is equal to the value of the floor. In this case, the energy company is said to be selling the collar.

FIGURE 3.7

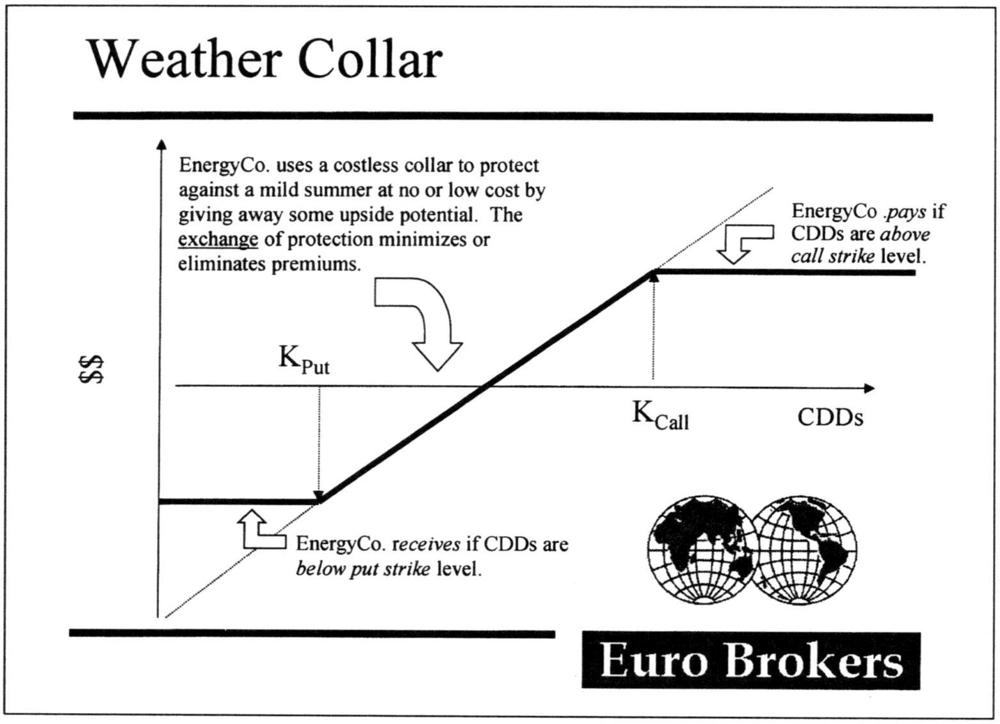

Customized derivative structures:

- **Baskets.** If a business has its weather exposure spread across multiple locations, then it is possible to enter into a derivative that pays out depending on the weather on each of these locations. This is called a basket. For example, if you want to protect against low HDDs over the winter in Boston, New York and Washington DC, and that you have twice the exposure in DC than you have in the other two cities then consider the following. First, construct an HDD index that is simply equal to 20% of the HDDs in Boston over the winter contract period, plus 20% of the HDDs in New York, plus 40% of the HDDs in Washington. Finally, enter into an HDD floor contract based on this index, which will pay out for a mild winter, with greater emphasis on the actual weather in Washington.

- **Digitals.** Whereas HDD and CDD caps, floors and swaps tend to have fluid pay out structures, i.e. the amount paid depends on how far the index deviates from the strike, a digital structure will make a single fixed lump sum payment simply if the strike level is breached. Because the nature of the pay out is all-or-nothing, these structures are also referred to as binary derivatives. In the example below, the power producer knows that there is a fixed cost of bringing a peaking facility on line when temperatures exceed 95 degrees Fahrenheit. The company buys a digital call option that will compensate him for this one off fixed cost when the daily maximum temperature exceeds 95 F. Of course, the buyer will pay the seller an up-front premium for this contract.

FIGURE 3.8

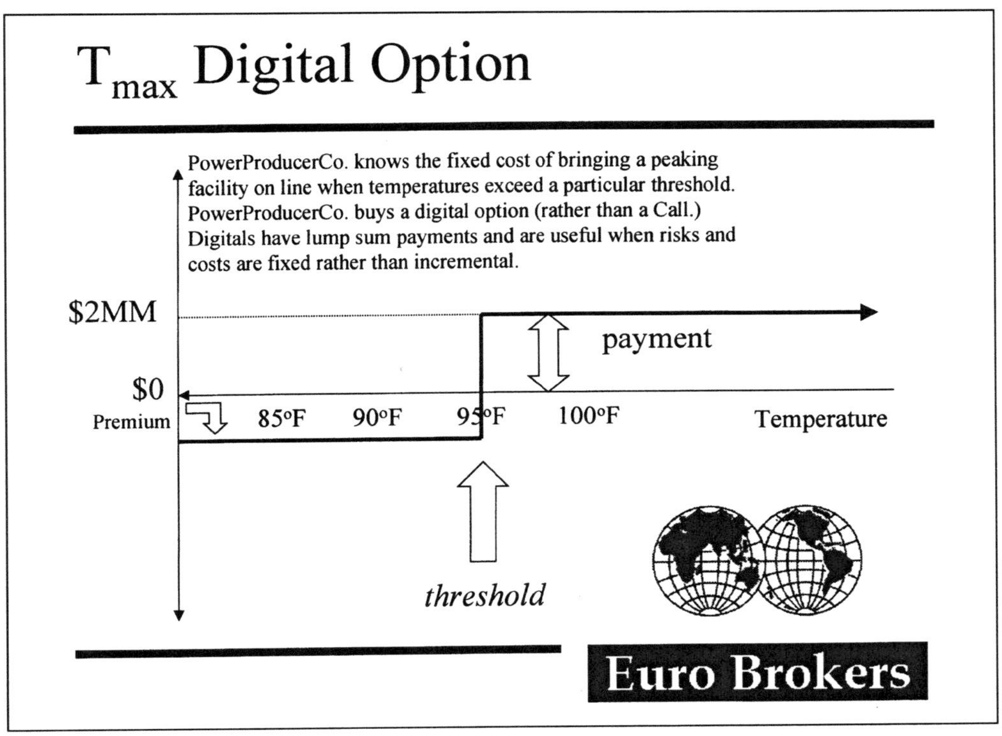

- **Compound Options.** Simply, an option on an option. If a business is unsure that it will need to take up weather risk protection it can purchase a contract which gives it the right, but not the obligation, to buy a specific weather option at a pre-determined price before a certain date. For example, a business will know if a winter (November to March) HDD floor option will best suit his risk hedging needs one month in to the contract period. He can buy a compound option on this structure, which expires on 31 December. So, if the regular floor costs $300,000, the company may expect to pay $75,000 for the compound and should they wish to exercise and take up the full November to March option on 31 December, they would pay an additional $275,000. It is reasonable to expect that the full cost of the exercised compound option be more expensive than the regular floor because of the greater flexibility it gives to the buyer.

- **Trigger Options.** Also known as barrier options, these structures either come into existence or cease to exist if the weather index breaches some pre-determined level. Knock-ins are options that come in to existence when this level is breached, knock-outs will cease to exist when this level is breached. Knock-outs have been more common to date. An example would be a November to March HDD call option with a strike of 5150. The option pays $10,000 for every HDD the index settles above the strike up to a maximum of $4,000,000, or 400 HDDs above the strike. If the option knocks-out at 5550 and if the index settles above this level, then instead of the buyer receiving the maximum four million dollars the option will expire worthless. Naturally these types of option will be less expensive than their non-trigger equivalents.

- **Dual Commodity.** A weather option that pays out in the form of another physical or derivative commodity contract. For example, a power distributor who wishes to receive a call option on electricity when temperatures are extremely high during the summer, in order to protect against price spikes, would consider a dual commodity structure.

CASE STUDIES

The following are examples of structures that either traded or were actively quoted.

1. Pittsburgh, PA:

Type of Derivative:	Summer CDD Floor
Weather Index:	CDDs
Location:	Pittsburgh International Airport, WBAN # 94823
Risk Period:	1 June, 1999 to 30 September, 1999
Strike Level:	630 CDDs
Pay Out Rate:	$10,000 per degree day
Limit:	$2,000,000

The buyer of this option is protecting against a summer with a low cooling degree days value. Low CDDs means a cool summer, leading to diminished air conditioning demand and therefore lower demand for electricity. Under this contract, the buyer receives $10,000 for every CDD the index finally settles below 630 at the end of the contract period. The maximum pay out is $2,000,000 which is achieved if the CDD index settles at, or below, 430.

2. Dual Commodity—Weather / Power Structure:

Structure:	Peak only, Monday to Friday
Risk Period:	1 June, 1999 - June 30, 1999

CASE STUDIES (continued)

Strike:	T_{max} Forecast ≥ 92 F at Cincinnati Covington Airport (North Kentucky), WBAN # 93814
Daily Power Call Strike:	$100
Power Settlement Index:	MegaWatt Daily Index for Spot Daily Electricity into Cinergy
Settlement:	Financial
Size:	50 MW/hr

This structure would be used to protect against peak power price spikes caused by high temperatures. Notice that the weather index is based on forecast temperatures. The reason for this is that prices for next day electricity trades are heavily dependent on forecast weather. The buyer of this contract will receive an electricity call option for each day that the forecast maximum temperature is greater than, or equal to, 92 F.

The option he receives will have a strike of $100 and settle against MegaWatt Daily's index for spot daily electricity at the Cinergy hub. So, for example, if the next day forecast temperature on Tuesday 8 June, 1999 is 95 F, the buyer of the option will receive the electricity call option for Wednesday 9 June. Suppose the MegaWatt Daily index on 9 June settles at $250 then the buyer will receive 250 - 100 = $150 for every MW in the peak day. There are 12 hours in the peak day and the size is 50 MW/hr, so he will receive 12x50x150 = $90,000.

In this case the power element was settled financially. It is good to bear in mind that it is possible to settle with physical delivery of electricity.

CASE STUDIES (continued)

3. Off-Shore Hurricane Structure:

Type of Derivative:	Daily Wind Speed
Weather Index:	Maximum sustained surface wind (mssw) as reported by the National Hurricane Center
Location:	92"20'W, 19"15'N
Risk Period:	1 May, 1999 - 31 July, 1999
Strikes:	In each 24 hour period, a Tropical Cyclone is reported to have mssw ≥ 25 knots and the storm centre is within the defined radius from the location
Pay Out Amounts:	$125,000/day if storm centre is within a 500km radius $62,500/day if storm centre is within a 750km radius
Limit:	$2,500,000

This structure was designed to protect an oceanographic survey project. The contractor knew that he would lose $125,000 each day in salaries, equipment rentals and overtime penalties if he had to leave the survey location and head to shore because of an impeding hurricane. This contract would pay him the full $125,000 for each day a hurricane came within 500km of his location off the Yucatan peninsula in the Gulf of Mexico. He would also receive a 50% pay out for each day a hurricane came within 750km as this would put him on alert and limit his activities. The cost of this contract was approximately $200,000.

CASE STUDIES (continued)

4. Euro Brokers Quote Sheet:

The following sheet contains real prices from last August for the 1999/00 winter season. The vast majority of the structures regularly quoted are degree day seasonal options. The sheet lists the location, the WBAN number for the location, the type of weather index (HDDs or CDDs), the risk period (or term), strike level, option type (put or call), the bid and offer in thousands of dollars, the pay out per degree day in thousands of dollars, the maximum pay out in millions and finally the maximum pay out in degree days. With the bids and offers for the compound options, the first number is the cost of the compound and the second number is the extra amount to pay should you choose to exercise the option on or before the expiration date.

TABLE 3.9

Euro Brokers Inc.

Weather Derivatives Price Sheet August 30, 1999

Telephone 203-602-2150

Ellen J. Slote
Andrew Feachem

SWAPS

	Location	WBAN#	Degree Days	Term	(* Subject) Bid	Offer			10 year average
1	Atlanta	13874	HDD	Nov-Mar	2405	2455	2.5	1	
2	Bismarck, ND	24011	HDD	Nov-Mar	6650	6740	5	2	
3	Boise, ID	24131	HDD	Nov-Mar		4380	2	0.8	
4	Central Park	94728	HDD	Nov-Mar	3800		5	2	
5	Charleston SC	13880	HDD	Nov-Mar	1575	1720	2.5	1	
6	Chicago	94846	HDD	Nov-Mar	5025		5	1	
7	Chicago	94846	HDD	Dec-Mar	4200	4350	5	1.5	
8	Cleveland	14820	HDD	Nov-Mar	4690	4725	5	1	
10	Dallas	03927	HDD	Nov-Mar	2105	2165	2.5	1	
11	Indianapolis	93819	HDD	Nov-Mar	4385	4475	5	2	
12	LaGuardia Arp	14732	CDD	Sept		185*	2.5	0.3	
13	Las Vegas	23169	HDD	Nov-Mar	1980		5	1	
14	Jacksonville, FL	13889	HDD	Nov-Mar		1250	5	2	
15	Memphis	13893	HDD	Nov-Mar	2625	2715	5	2	
16	Portland, Or	24229	HDD	Nov-Mar	3155	3190	3	1.2	
17	Seattle	24233	HDD	Nov-Mar	3190		2	0.5	
18	Spokane	24157	HDD	Nov-Mar	4815	4870	5	2	
19	Sacramento, CA	23232	HDD	Oct-Dec	1005		2.5	1	
20	San Francisco	23234	HDD	Nov-Mar	1840	1925	5	1	
21	Sioux Falls S.D.	14944	HDD	Nov-Mar	6040		5	2	
22	Tulsa	13968	HDD	Nov-Mar	3140	3180	5	1	
23	Wichita, KS	03928	HDD	Nov-Mar	3910	4040	5	2	

This information is proprietary and confidential

PRICING ISSUES

There has not been much published on this topic and, to be honest, this topic deserves an entire book let alone a segment in a chapter. So I will just touch on the key issues.

Since there is no underlying index to base prices on, it is inappropriate to use the traditional Black Scholes option model. Black Scholes also assumes a normal distribution in index movements and this is weak assumption to make for temperature movements.

The simplest method employed has been the use of *burn costs*. You calculate how the contract you are pricing would have paid out historically and then take the average of the payouts. This is very basic and does not take into consideration that there is likely to be some kind of temperature trend over the years. It is also entirely subjective as to which historical time period you take the average from. It makes more sense to look at the 15-year average rather than the 30-year average as data trends will distort the 30 year value.

Another method is known as *bootstrapping*. Taking data over the risk period from the previous 10 years, you randomly generate temperature sequences from the real observed historical data. You do this 10,000 times and observe the pattern of resulting payouts. For example, if pricing a June option, the random generator might take the temperature from 1 June, 1993 followed by the temperature from 2 June, 1997 followed by the temperature from 3 June, 1990 and so on, until a temperature sequence for the entire month is created. Doing this 10,000 times will not only give you an average value, but also a distribution on which to base probability of various pay out calculations. Plus, taking the past ten years lessens the need to de-trend the data whilst giving you 10,000 payouts on which to base your calculations.

Possibly the most rigorous method would be a *Monte Carlo simulation*. First you would de-trend the historical data. Secondly you must construct a model to simulate future temperature patterns. The model will inherently draw on econometric methods. It might incorporate a mean reversion factor, which will always pull back simulated temperatures towards some notional historical average. As in interest rate models, this prevents the possibility of simulating 1000 F temperatures in New York over the winter. The model might also have factors that state that today's temperature is very highly correlated to yesterday's temperature. It is also reasonable to assume that the variance and average of temperatures during June will be different than for August. Once the model has been constructed, parameters for the various factors must be estimated. These parameters will most certainly be different from location to location and is the most computationally intensive part of the procedure. Based on this, the model is run several thousand times in order to create a distribution for temperatures over the risk period on which to calculate the value of the derivative.

SUMMARY AND OUTLOOK: MONTE CARLO OR BUST?

From the inception of the weather derivatives market just two and a half years ago, it is now estimated to be worth in excess of $5bn with the vast majority of trades being in the secondary market. They are increasingly becoming standard fare in the portfolios of utility companies with non-US based deals set to be of increasing importance over the coming year.

The growing US market in weather products sparked the beginning of the European market, with the first deals being completed for the winter of 1998. The market on the continent, however, is developing in a different way. So far, all the deals have been transacted with end-users and the secondary market looks still to be some months away. That is not to say that the response by European companies to the use of weather products has been anything but positive. The European market will grow rapidly over the coming year as companies see for the first time other players within their industry who have successfully completed a weather hedge over the current winter season. Liquidity will build up over the major hubs—London, Paris, Berlin, Oslo etc.- allowing easier access to the weather market to new participants.

One further difference for the European market is that risk hedgers are more interested in absolute temperature contracts, digital payouts and precipitation contracts. Not the plain vanilla HDD and CDD options and swaps that have become the mainstay of the US market. These are the types of structure that seem to best offset the utility's weather exposure and demand for these structures is now growing back in the US within the energy industry. A reason for this is that when the weather negatively affects an energy company, it will often affect its profits in one large hit, regardless of how extreme the weather event was.

The nature of weather derivatives gives brokers a valuable role in the development of the market. The products are still relatively illiquid, certainly when compared to the money markets, there is little transparency to market activity and the business is notorious for its lack of published information. Brokers provide much needed market information, such as where contracts have traded in the past, assistance with historical weather data, price discovery and assistance on structuring deals.

This year will be make-or-break for the global weather derivatives market. In the US, there will be growing demand for digital structures and contracts based on precipitation and absolute temperatures. Europe will see far greater activity and East Asia and South America will become active markets. Increased participation from the insurance and reinsurance industry will lend far greater capacity to the market and we will begin to see more activity from the non-energy sectors, such as the beverage and entertainment sectors.

It looks like the weather derivative market is going to make it to Monte Carlo.

CHAPTER 4

Emissions: Trading in Practice

By Garth Edward and Matthew Varilek

This chapter assesses the experience of emissions trading programs and the lessons that can be drawn for companies as they consider their competitive position under a carbon constrained future.

EMISSIONS TRADING VERSUS TRADITIONAL REGULATION

Traditional regulatory approaches to controlling emissions mandate installation of specific emissions reduction technologies or impose specific emissions reduction requirements. These "command and control" approaches face increasing criticism from affected companies protesting the prohibitive costs of carrying out these mandatory actions. Opponents claim that it is inefficient to require uniform quantities of emissions abatement by sources whose marginal costs of abatement may differ widely.

Emissions trading responds to these criticisms by devolving authority for choosing the method and location of emissions reductions to businesses. Several advantages have been demonstrated:

- **Secure Environmental Results.** Command-and-control air quality regulations usually aim to reduce emissions by uniformly limiting individual sources' allowed rate of emissions per unit of energy input or mandate installation of a particular emissions control technology. Fiscal instruments such as emissions taxes charge sources per unit of pollution emitted or energy consumed. While these methods of regulation can be effective in restricting emissions, their exact environmental results are hard to predict since they set no aggregate emissions limit. Per unit emissions reductions may be offset by overall growth in production. The result of a per unit charge for emissions cannot be predicted without knowing the elasticity of demand for the good whose production creates emissions. In contrast, emissions trading can achieve precise environmental objectives. By distributing a finite quantity of emissions rights amongst a set of affected sources, administrators can assure that actual emissions from those sources will be within an agreed aggregate limit.

- **Lower system-wide compliance costs.** Emissions sources are allowed to achieve a given aggregate emissions reduction level by carrying out the most affordable reduction opportunities, rather than ones specified by government. This flexibility can produce substantial cost savings relative to a command-and-control approach. For example, the United States General Accounting Office estimates aggregate annual compliance costs for 2002 of US$1.4 billion for its flexible Sulfur Dioxide (SO_2) allowance trading system, as compared to estimated costs of US$4.5 billion to achieve the same reduction under a traditional regulatory approach.[1]

- **Lower individual compliance costs.** Under a command-and-control regime, those companies whose abatement costs are highest bear a disproportionate share of the overall compliance burden. Emissions trading spreads this burden by providing high-cost abaters with a cheaper means to compliance and low-cost abaters with an opportunity to earn profits. Rather than undertake all its required emissions reductions within its own business units, a high-cost abater might pay a low-cost abater to exceed its required reductions. In return, the high-cost abater would receive legal rights to claim the additional reductions as its own. The high-cost abater would benefit as long as it could buy emissions reduction rights more cheaply than it could reduce its own emissions. The low-cost abater would benefit as long as it could sell the rights to its additional reductions at a price higher than the cost of achieving them. In the US SO_2 market, for example, individual American utilities such as Duke Power and Central Illinois Public service reported annual savings US$300 million and US$225 million respectively compared to anticipat-

[1] US GAO (1994)

ed compliance costs under a command-and-control regime.[2] These so called "gains from trade" can potentially be realized, to the advantage of both transacting parties, wherever marginal abatement costs differ.

- **Positive incentives for efficiency and innovation.** Whereas command-and-control regulations encourage businesses only to achieve minimum compliance levels, emissions trading creates positive incentives to exceed required reductions. The potential to earn revenue from the sale of excess emissions reductions encourages businesses to undertake a variety of desirable activities such as improving energy efficiency, switching to cleaner fuels, and developing new emissions-reducing technologies. Savings from these adjustments can further reduce the overall costs of compliance with environmental regulations.

DESIGN COMPONENTS

As the advantages of emissions trading become more apparent, use of this tool as an alternative to traditional regulatory methods will continue to grow. In order for companies to optimize their performance within this new regulatory context, they must understand the basic building blocks of emission trading programs.

- **Market structure** refers to the regulatory framework in which emission permits would be tradable. For example, in a **cap-and-trade** system, the regulatory authority would impose an absolute ceiling (or "cap") on the aggregate emissions from affected sources. Those sources would receive an allocation of tradable allowances that authorize emissions of a portion of the aggregate ceiling. A **baseline-and-credit** system would impose individual emissions restrictions on affected sources. Those whose emissions were less than the quantity allowed by their individual restriction would earn credits that could be traded to sources whose emissions exceeded their allowed quantity. A hybrid would apply different structures to different sectors.

- **Unit of Exchange** refers to the commodity to be traded, such as "allowance to emit one ton of CO_2 equivalent during 2008". **Banking** concerns the duration over which a permit is valid as a license to emit.

[2] Guerrero (1997)

The term derives from the notion that a participant could potentially save or "bank" some unused credits from a particular compliance period for use in a future compliance period.

- **Participation** concerns the range of sources who will be regulated by the system, and the extent of their regulation.
- **Allocation** refers to the month by which emissions permits will be granted to regulated sources.
- **Monitoring** refers to tracking of actual emissions levels as well as transfers of permits.
- **Verification** refers to confirmation by an outside party such as an accountant or engineer that a claimed emissions reduction has occurred. **Certification** refers to the conferral of credits by the trading authority for claimed emissions reductions relative to an agreed baseline. Certification requires not only that a verified reduction has taken place, but also that the nature of the reduction conforms to rules stipulating what reduction activities are permissible.
- **Enforcement** concerns measures that will be taken to ensure compliance with the rules of the trading system.
- **Method of exchange** refers to the means by which permits will be traded such as Over-the-Counter (OTC) brokerages, formal exchanges, or screen-based exchanges.
- **Timescale** refers to the phasing in of compliance requirements.

A REVIEW OF EMISSIONS TRADING IN PRACTICE

The following section reviews the experience in US emissions trading programs at both the state and federal levels, including the: (1) US Sulfur Dioxide Allowance Program, (2) US NO_X Ozone Transport Commission Allowance Program, (3) US Open Market Trading Program for Ozone Smog, and (4) Precursors Southern California Regional Clean Air Incentives Market (RECLAIM).

(1) US Sulfur Dioxide Allowance Program

The United States' Sulfur Dioxide (SO_2) Allowance Program is the most sophisticated and mature of the world's permit markets. It was established by a provision in Title IV of the 1990 Clean Air Act Amendments to reduce industrial emissions of SO_2, which contribute to the creation of acid rain. The program's legislative basis is unique amongst environmental programs in that a federal agency administrates and enforces the program. While it is common for the federal government to issue environmental standards and rules for meeting the standards, individual states are usually responsible for administration and enforcement. The SO_2 allowance program's success is often cited as an example of the potential for emissions trading to achieve high rates of compliance with stringent environmental goals at low overall cost to the economy.

- Description

Design

1. *Market Structure.* The program's eventual goal is to achieve an overall cap of 8.95 million tons (Mt) of SO_2 emissions from utilities by the year 2010. This represents a reduction of roughly 50% relative to 1980 emissions levels. Affected sources receive an allocation of tradable allowances. At the end of every year, the emissions of each affected source must be less than or equal to the quantity of emissions permitted by its holdings of allowances.

   During a "true-up" period from January 1 of the year after a given compliance period until the end of February in the same year, sources prepare and submit records detailing their emissions and holdings.

Sources are allowed to trade during this period in order to cover any discrepancies between planned and actual emissions. On March 1, the market's regulatory authority within the Environmental Protection Agency (EPA) freezes all accounts that track each source's holdings of allowances. If a source's emissions exceed its allocation, it would be subject to non-compliance provisions. EPA cancels the validity of allowances that have been submitted to cover actual emissions so that they may not be reused in the future. Surplus allowances may be banked for future use (see 2. *Unit of Exchange* below).

The program is widely regarded as the most successful example of a pure cap-and-trade system. However, it is important to note that it includes a provision that allows sources to earn allowances for certain project-based activities (see 6. *Certification and Verification* below), a feature typically associated with baseline-credit systems.

2. *Unit of Exchange.* The tradable commodity created by the Title IV SO_2 provision is known as an "SO_2 emissions allowance". Each allowance entitles its holder to emit one ton of SO_2. An allowance is legally classified as an "authorization to emit" rather than a "property right", which implies that the regulatory authority has the power to rescind the authorization. However, the authority is unlikely to do so because of the legislative difficulty and likely political repercussions of such an action. In spite of this legal subtlety, allowances are freely traded as if they were property rights. Confidence in the long-term viability of the program is aided by EPA's publication of 30-year allocation schedules for each source, which provides a basis on which to formulate future operational plans.

   Each allowance carries a unique serial number by which it can be identified over the course of its existence and which indicates its year of origination or "vintage". Vintage is only important for tracking purposes, not for regulation, as there are no temporal limits on trading. This means that any quantity of allowances may be banked at full value for future compliance periods. Trading is also free of geographic restrictions even though SO_2 does cause some localized impacts. Rather than complicate the market with geographic trading restrictions, EPA left individual states the authority to address local concerns. So while a source may achieve compliance with the federal SO_2 program by obtaining adequate allowances, it must also comply with local emissions regulations that vary from state to state.

3. *Participation.* The range of sources affected by the program and the degree of their obligation changes over the course of two phases (see also 10. *Timescale* below). Phase I, from the beginning of 1995 to the end of 1999, required mandatory participation by 263 of the country's largest SO_2 emitters. These consisted primarily of coal-fired electric utilities with emissions greater than 2.5 pounds of SO_2 per million Btus of fuel consumed, and with a nameplate capacity of at least 100 Megawatts.

   Phase I also allowed for voluntary participation by other utilities that would not face mandatory regulation until the beginning of Phase II in 2000. Upon their voluntary accession to Phase I, these units faced adherence to the same rules as the original affected sources. Phase II units that voluntarily joined Phase I probably calculated that their potential to reduce emissions would allow them to emit considerably less than their anticipated allocation of allowances. Taking advantage of these early reduction opportunities would allow the sources to withhold a surplus of allowances that could either be saved for use against their own future compliance requirements, or sold for financial gain. 182 units voluntary joined Phase II, bringing the total Phase I participation to 445 units. However, because there are no restrictions on who may trade (i.e. trading rights are not limited to affected sources), and because individual market participants may trade on behalf of more than one affected unit, the actual number of active market participants is difficult to determine. At any one time during Phase I the number of participants ranged from 100 to 250.[3]

   Phase II requires mandatory participation by all new utilities of any size and existing utilities that burn fossil fuels with at least 0.05% sulphur content and have at least 25 Megawatts of generating capacity. This will amount to around 2050 sources. These will account for roughly 70 percent of all SO_2 emissions in the United States. The program does not apply to other emissions sources such as industrial facilities or automobiles.

   Besides the provision that allows early participation by Phase II sources, another provision allows voluntary participation by emitters of SO_2 who would not otherwise be required to participate in either phase.

[3] Natsource interview, February 2, 2000.

Sources that chose to "opt-in" under this provision would have received a grandfathered allocation of allowances and faced the same rules of participation as other sources. The provision was intended to expand the percentage of national SO_2 emissions covered by the program, and to take advantage of cheap emissions reduction opportunities in the industrial sector and at utilities not large enough to face eventual mandatory regulation. However, in practice very few non-affected sources exercised this option. The high cost of installing required monitoring systems (see 5. *Monitoring* below) apparently deterred some potential participants.

Since the program applies to neither to suppliers of fuel whose combustion produces SO_2 nor to the consumers of the final product of that combustion, it is best described as a "mid-stream" point of application. Whereas an upstream approach would have been simpler to administrate because it would have had fewer participants, a more complex downstream approach would likely have produced greater liquidity. The midstream choice represents a compromise between these concerns about breadth of participation and administrative complexity.

4. *Allocation.* Affected sources that were in operation before 1995 receive a grandfathered annual allocation of allowances at no cost. The size of each allocation is calculated on the basis of a numerical rule that sets a standard allowable rate of SO_2 emission per unit of fuel consumed. The program's overall cap was reduced by a change in the basic allocation rule that took effect at the beginning of Phase II. Besides these standard allocations, provisions in Title IV include exceptions for individual units and offer "bonus" allowances and such that the total allocation of allowances deviate from that predicted by the basic rule. EPA also reserves a quantity of allowances that are offered for sale on an annual auction.

   Phase I units received an allocation of 2.5 pounds of SO_2 per million Btus of heat input multiplied by the unit's average heat input over the base years of 1985-7.[4] Affected units that began operation after the base years but before 1995 received an allocation according to the same for-

[4] A complete list of Phase I allocations can be found at www.epa.gov/acidrain/lawsregs/p1table.html

mula but on the basis of a mutually agreeable base year. Units that began operation after 1995 received no allocation and were forced to acquire allowances on the market. In Phase II, the basic rule changed to allocate only 1.2 pounds of SO_2 per million Btus of heat input. When the total allocation according to this rule exceeds the program's overall cap of 8.95Mt of SO_2 in Phase II, each source's allocation is "ratcheted" down proportionally so that the cap is met.

Various provisions in the Title IV statute create additional allowances in both phases, though in Phase II, these must not cause the overall allocation to exceed the cap. Individual units were granted exceptions for special circumstances such as lower-than-normal capacity utilization or demand during the base years. These exceptions were won through lawsuits against the EPA.[5] Some states were granted a discretionary budget of additional allowances that they could access in order to offset their perceived disproportionate compliance burdens of their state's units. Key congressional officials who participated in drafting the acid rain legislation influenced allocation of these budgets.[6] A reserve of allowances was also created to reward units that installed flue gas desulfurization systems (scrubbers). Some have described this provision as a concession to states that produce high-sulfur coal, which fuels many of the affected utilities. The bonus allowances may have been intended to encourage units to achieve compliance with scrubbers instead of switching to a low sulfur coal.[7] Additional allowances could be earned for certain "early activities" undertaken by source before they faced mandatory participation in the program (see 6. *Certification and Verification*).

Each year the EPA auctions 2.8% of the total annual allocation on the Chicago Board of Trade. The allowances auctioned are culled proportionally from the allocations to affected sources, and revenues from the auction are refunded to those sources in the same proportion. The auction offers assured market access for new sources and other market participants, and provided a valuable price signal in the early days of trading. As market liquidity has improved, however, the importance of this latter function has declined.

---

[5] Natsource interview, February 2, 2000.
[6] Joskow and Schmalensee (1997)
[7] ibid.

5. *Monitoring.* Affected sources are required by EPA to install a Continuous Emissions Monitoring system (CEM) to measure and record their emissions. Sources compile and submit quarterly reports that detail their hourly emissions rates as measured by the CEM. The reports are recorded by EPA on an Emissions Tracking System (ETS), and are used by EPA to assure compliance with the program by comparing sources' actual emissions with their holdings of allowances. Data on the ETS are publicly available, serving as evidence of the program's environmental effectiveness.

   Allowances exist only in electronic form. EPA monitors allowances on an electronic registry called an Allowance Tracking System (ATS). Each source, as well as any other holder of an allowance, is assigned an account that records the information about the participant's holding of allowances. In addition, the system records information about transfers of allowances, deductions for compliance purposes, and holdings of allowances in the various reserves. This information is available to public on the internet. The system does not track price data. EPA explains that "such information is better collected and reported by the private sector through established exchanges or other trade information brokers".[8] (see Section B on "The Role of Information Technology" for an additional discussion of SO2 monitoring provisions.)

6. *Certification and Verification.* Submission of quarterly emissions reports as described above fulfills certification and verification functions in most cases. For emissions during a compliance period, no additional certification or verification is required except for regular maintenance of monitoring equipment by authorized technicians.

   The program includes two provisions that allow sources to earn allowances for activities undertaken between the date when rules where promulgated and before the source's date for mandatory participation. The first is a reserve of 300,000 allowances to reward utilities for approved investments in demand-side energy efficiency projects or renewable energy generation projects. For example, a Phase I utility would have earned bankable credits for a renewable energy investment

[8] US EPA (1998) www.epa.gov/acidrain/ats/atsintro.html

undertaken in 1993. Only 34,638 of these allowances were awarded.[9] Another provision allows utilities of certain narrowly prescribed characteristics to earn allowances for early activities related to "physical changes or changes in the method of operation made after the date of enactment of the Clean Air Act Amendments of 1990, including changes in the type or quality of fossil fuel consumed".[10] This provision was included to appease one particular utility, and it was later accessed by another utility that won a suit against the EPA claiming that it too met the conditions for certification. 629,438 allowances were granted under this provision. Reductions undertaken for both this provision and the conservation and renewable energy reserve were verified by state public utilities commissioners or by independent auditers.

7. *Enforcement.* Sources are out of compliance if at the end of the true-up period their actual emissions in the past year are shown to have exceeded the quantity authorized by their allowances. An automatic penalty of $2,000 per ton exceeded (indexed to inflation from 1995) is assessed, and an allowance for every ton exceeded is subtracted from the violator's current allocation as recorded on the ATS.

   ATS account holders may petition EPA for review of decisions related to allowance transfers, deductions, or sales. If the petition relates to an accounting error on the part of EPA, the petitioner must first file a claim of error.

8. *Supplementarity.* The SO_2 allowance program is not linked to any external system that could provide participants with access to additional credits. Neither are there provisions to allow for interpollutant swaps for compliance, though participants are not prohibited from swapping SO_2 allowances for other types of allowances in bilateral trades.

9. *Method of Exchange.* Except for its annual auction, the EPA does not regulate the mechanism by which allowance trades occur (see Section B on "The Role of Information Technology" for an additional discussion of SO_2 trading activity).

---

[9] US EPA (1999) http://www.epa.gov/acidrain/crer/crertbl.htm

[10] 1990 Clean Air Act Amendments, Section 404, paragraph 'e'.

10. *Timescale.* Because the SO_2 allowance program was intended to operate independently, its implementation was not coordinated with other regulatory systems. Most operational rules such as those concerning trading, monitoring requirements, allocation, non-compliance penalties were established and promulgated in March of 1993, almost two years prior to the first compliance period beginning in 1995. The ATS became operational in March 1994.

    As explained above, coverage of participants and the quantity of allocations were phased over time. EPA sought to maximize the program's immediate environmental impact with a minimum of administrative complexity by focusing Phase I on the largest and most intensive SO_2 emitters.

(2) US NO_X Ozone Transport Commission Allowance Program

The NO_X Ozone Transport Commission (OTC) Allowance Program is relatively new permit market designed to control emissions of oxides of nitrogen (NO_X) in the northeastern United States. NO_X is a precursor of ground-level ozone, and it contributes to formation of acid rain. The NO_X allowance program is modeled on EPA's successful SO_2 allowance program.

The program was initiated by a 1994 Memorandum of Understanding[11] (MOU) between eleven northeastern states and the District of Columbia who collectively comprise the OTC.[12] The MOU committed its signatories to develop a regional emissions trading system in cooperation with the US EPA. After agreeing rules for operation of the trading system, each OTC member was responsible for passing regulations to implement the program at the state level. The EPA continues to play an important role by maintaining electronic registries of emissions data and allowance transfer activity, as well as conducting an annual assessment of whether sources' allowance holdings are sufficient to account for their emissions. However, the program's ultimate regulatory authority rests with each of the participating states.

[11] Ozone Transport Commission (1994)

[12] Virginia is also an OTC member, though it did not sign the MOU and does not participate in the emissions trading market. The other members are Maine, New Hampshire, Vermont, Massachusetts, Connecticut, Rhode Island, New York, New Jersey, Pennsylvania, Maryland, and Delaware

Since the program's first compliance period was in 1999, it is still too early to draw definitive conclusions about the program's performance. Several states did not participate in the first year for various reasons, and regulatory developments were perhaps as important a driver of market behavior as compliance costs. Nevertheless, the accumulation of a large bank of unused allowances suggests that affected sources have identified cost-effective ways to meet and exceed the program's environmental objectives.

■ Description

Design

1. *Market Structure.* The program presently caps annual emissions of NO_X from affected sources at 218,367 tons. In 2003 the cap will fall to 142,874 tons per year, which represents a 75% reduction relative to 1990 levels.

   While the MOU specified the amount of this intended emissions reduction, it did not mandate a particular type of market structure. One regulator explained that state signatories to the agreement chose a cap-and-trade system in order to achieve more precise aggregate environmental results than an alternate structure might have provided. Citing excess production capacity in the electricity sector, and annual capacity growth of roughly 2% per year, the regulator expressed concern that environmental gains achieved by a baseline-and-credit system with rate-based emissions limits might have been overwhelmed by overall increases in production.[13]

   Affected sources receive an annual allocation of tradable allowances. At the end of each calendar year, sources must demonstrate that their emissions during the "ozone season", which lasts from May through the end of September, were no greater than the quantity of emissions authorized by their holdings of allowances. The ozone season is the period in which ground-level ozone problems tend to be most acute, due to reactions of NO_X and Volatile Organic Compounds (VOCs) with sunlight.

[13] Natsource interview, February 23, 2000

Though the compliance period for emissions is restricted to the summer months, trading continues throughout the year. During a "true-up" period, which lasts from the end of the ozone season until the beginning of the next year, sources have an opportunity to check their records of emissions and allowance holdings for accuracy, and to reconcile any differences by trading allowances. Sources whose emissions exceed their allowances at the end of the true-up would be subject to non-compliance provisions. At the end of the year, EPA cancels the validity of allowances that have been used to cover actual emissions so that they are not reused in the future. Unused allowances may be banked for future use subject to restrictions (see 2. *Unit of Exchange* below).

2. *Unit of Exchange.* The tradable commodity is known as a "NO_X" allowance. Each one entitles its holder to emit one ton of NO_X during the ozone season of a given year. Every allowance carries a unique serial number, which allows it to be tracked and identified by its year of origination or vintage. Vintage is necessary to calculate the value at which unused allowances may be banked for future use. Banking is governed by a system of restrictions called "progressive flow control". It says that if the quantity of unused allowances from a given year is less than 10% of the program cap, allowances are banked at full value. If the quantity of unused allowances exceeds 10%, allowances are banked at a discounted rate.[14]

   Though NO_X does cause localized impacts, trading is not geographically restricted within the OTC. As a safeguard against "hot-spots", or locations with high emissions concentrations, sources remain subject to other federal and state regulations such as the Reasonably Available Control Technology (RACT) in Title I of the 1990 Clean Air Act Amendments. Since these regulations generally permit higher emissions levels than a source's allowance allocation, they would only come into effect if the source were a significant net buyer of allowances.

3. *Participation.* The program applies primarily to large utilities and industrial sources of NO_X within the OTC. Nationwide, roughly 33% of NO_X

[14] For a full explanation of the formula used to calculate the discount, see NESCAUM/MARAMA (1996).

emissions come from utilities, and 13% from the industrial sector.[15] Participation is required of fuel boilers or indirect heat exchangers with a maximum rated heat input capacity of 250 MMBtu per hour or more and electricity generating utilities of at least 15 MW of rated output. Roughly 450 sources fall into these categories. These account for roughly 75% of emissions from utilities and industrial sources within affected states.[16]

Because a single NO_X trader may be responsible for several sources, the number of market participants is considerably less than the number of affected sources. One broker placed the number of traders at around 100, and the number who are active on a daily basis at around 12.[17] However, these figures vary throughout the year since trading activity, like the compliance period, is seasonal.

Most descriptions of the program note that Virginia, as a non-signatory of the 1994 MOU, elected not to participate in the trading program. This implies, however, that the remaining members of the OTC did participate. In fact, only eight of the twelve OTC states participated in the program's first year, and only seven of them traded allowances. Utilities in Maryland won a state lawsuit that exempted them from compliance. The District of Columbia, with only two major sources, failed to adopt a rule implementing the program. Vermont did not receive an allocation of allowances since it has no major emissions sources. Maine, despite signing the MOU, chose to continue regulating its utilities with command-and-control regulations. Rhode Island participated to the extent that it received an allocation of allowances, but it prohibited its sources from inter-state trading.

Other sources within the OTC that do not face mandatory participation may opt in to the program. They would receive an allocation of allowances based on their 1990 emissions, provided that this would not provide them with a built-in surplus. Once acceded to the program, voluntary sources may not opt out unless they cease operation.

Any person may acquire and trade allowances.

---

[15] Farrell *et al.* (1999)
[16] Natsource interview, February 22, 2000
[17] Natsource interview, February 14, 2000.

4. *Allocation.* The regional cap is distributed amongst its constituent states according to a geographically-differentiated formula. The MOU divides the states into three zones according to pollution levels, and allocates allowances to each state based on formula that applies to its zone. The most-polluted inner zone states face the most stringent reduction requirements, each receiving allowances worth only 35% of its 1990 emissions (or .2lb NO_X/MMBtu) for the first compliance period. The outer zone states receive 45% of their 1990 emissions (or .2lb NO_X/MMBtu). The northern zone states receive allowances equal to their 1995 emissions (RACT equivalent).

   As indicated above (see 1. *Market Structure*), during the first phase of the trading program[18] from 1999 to 2002, the regional cap is set at 218,367 tons per year. In 2003 the cap is set to decline to 142,874 tons per year. To achieve this reduction, zone allocations are reduced to 25%, 25%, and 45% of 1990 levels for the inner, outer, and northern zones respectively. However, federal regulatory changes set to occur in 2003 may supersede these designations (see 10. *Timescale*).

   Individual sources receive their allocations from the environmental regulatory authority in their state. Each state was allowed to determine its own method of allocation. Most states grandfathered allowances to existing sources on the basis of their historical fuel consumption. Sources could also have obtained allowances for certified emissions reductions achieved prior to the beginning of the program (see 6. *Certification and Verification* below).

5. *Monitoring.* Affected NO_X sources that are also participants in the US SO_2 Allowance Program are required to monitor their NO_X emissions with the same Continuous Emissions Monitoring Systems (CEMs) that monitor their SO_2 emissions. Other sources may use several approved alternative methods of emissions monitoring. Some monitor emissions directly by periodic stack tests. Others estimate emissions based on heat input or fuel consumption. All methods yield hourly emissions data.

   Reports are submitted electronically to the EPA, where the data are compiled and published on an electronic registry. EPA compares the

[18] Some literature refers to this period as Phase II, since the trading program follows an earlier set of NO_X regulations that constituted Phase I.

data with the allowance holdings of affected sources to determine their compliance status.

Allowances exist only in electronic form, and are tracked by serial number on EPA's NO_X Allowance Tracking System (NATS). The system records information about allocations of allowances, holdings, deductions of allowances for compliance purposes, and transfers between accounts.

Though NO_X is regulated primarily for its contribution to ground-level ozone, both the allowance tracking system and the registry of emissions data reside within the Acid Rain program website to save administrative costs.[19] Administrators determined that it would be cheaper to expand the existing SO_2 monitoring systems than to create an entirely new one.

6. *Certification and Verification.* The emissions monitoring systems described above fulfill the certification and verification functions for emissions reductions that occur during compliance periods. Sources also could have earned allowances for certified emissions reductions that occurred during the ozone seasons of 1997 and 1998. The baselines used to measure these reductions were calculated by applying the percent allocation of its state's zone to the source's own 1990 emissions level. For example, an inner zone source would have measured its reductions against baseline equal to 35% of its 1990 emissions. Early credits could not be earned for reducing capacity utilization. In order to earn allowances, early reductions had to be certified by the state regulatory authority. 24,171 allowances were granted for early reductions.

7. *Enforcement.* Sources whose ozone season emissions exceed their allowance holdings are out of compliance and subject to penalties. For every excess ton of emissions, a penalty of three allowances is deducted from the next year's allocation. Additionally, discretionary fines may be assessed. While EPA is responsible for determining whether sources are in compliance, the authority for assessing penalties rests with each state. Disputes are also addressed according to the normal administrative procedure of each state.

[19] See www.epa.gov/acidrain/otc/otcmain.html

8. *Supplementarity.* Except for its monitoring provisions, the NO_X OTC program is not currently linked to any other system that could provide participants with access to additional credits. However, the system may be expanded in 2002 to include an additional 10 states upwind of the current 12 (see 10. *Implementation*). Also, Rhode Island's trading rule is analogous to a supplementarity rule in that it restricts (i.e. prohibits) trading beyond the regulatory authority's jurisdiction.

9. *Method of Exchange.* States do not regulate the mechanism by which allowance trades occur.

10. *Timescale.* The states within the OTC are part of a larger group of 22 states that may be required by the federal government to implement a broader trading program (known as the "NO_X SIP call") in 2003. The existing trading program would have to be modified in potentially significant ways if the new regulation takes effect. Despite this possibility, the OTC states decided that the severity of their ground-level ozone problems warranted voluntary regional cooperation in advance of any federally mandated program. If the new regulations do not take effect, the existing program will continue to operate in its current form as specified in the MOU and subsequent regulations. The fate of the SIP call depends on the outcome of several pending lawsuits brought against the EPA by potentially affected states and industry groups that are opposed to the new regulation.

(3) US Open Market Trading Program for Ozone Smog Precursors

The US Open Market Trading Program for Ozone Smog Precursors was initiated in 1995 to control emissions of Oxides of Nitrogen (NO_X) and Volatile Organic Compounds (VOCs) from a broad range of sources. NO_X and VOCs are precursors of ground-level ozone, which causes respiratory problems and is a major component of smog.

Like the NO_X OTC program, the Open Market program is administered by individual states for the purpose of meeting National Ambient Air Quality Standards (NAAQS), which are federally mandated in amendments to the Clean Air Act. In order to foster interstate compatibility, EPA issued a model rule to guide each state's implementing regulations.[20] Main features of the model rule were agreed through discussion and negotiation amongst a variety of stakeholders including environmental, regulatory, and industry representatives. Despite the model rule and the collaborative process that produced it, implementing regulations vary considerably from state to state. Partially because of this variance, and more importantly because the regulated pollutants have primarily local impacts, interstate trading is uncommon. With a few exceptions, participating states operate what are, in effect, independently functioning markets. The following section describes and analyzes these markets collectively, focusing on their common features, while also mentioning some important differences.

Interstate incompatibility, while perhaps justified by individual circumstances and local pollutant impacts, has resulted in illiquid markets. Though participation is allowed by almost any emissions source, only stationary sources have shown significant interest in the program. With each market's group of potentially active participants thus restricted to stationary sources within the state, bids and offers for permits have been scarce. Consequently, transactions are rare, and few benefits have been realized as a result of the program. Even so, the program may be regarded as a success since it sometimes provides sources with a broader range of compliance options than would otherwise exist.

■ Description

Design

1. *Market Structure.* Open markets for ozone precursors are baseline-and-credit systems. They are described as "open" to distinguish them from "closed" cap-and-trade systems in which a finite quantity of tradable allowances exists. Open markets place no limit on the number of credits that may be generated (though they may place other restrictions on the type and duration of activities that may qualify for credit generation).

   To participate in the program, a source must adopt an individual NO_X or VOC emissions limit, known as a baseline, that serves as a refer-

[20] EPA (1995b)

ence point for calculating generation and consumption of emissions credits. Baselines for stationary sources such as utilities and large industrial facilities are usually mandated in existing command-and-control regulations, such as the Reasonably Available Control Technology (RACT) standard in Title I of the 1990 Clean Air Act Amendments. Other sources may voluntarily adopt a baseline in order to participate.

A source that emits less than the quantity authorized by its baseline could earn tradable credits for the difference between its actual and authorized emissions. Other sources could attempt to acquire those credits for use as legal compensation against emissions in excess of their own baselines. A source would be out of compliance if, at the end of its compliance period, its actual emissions minus its holding of credits were not within its baseline.

2. *Unit of Exchange.* Credits in the program may only be generated in discrete quantities for activities that create temporary emissions reductions such as fuel switching or operational changes. Generally, they can only be used to compensate for temporary periods in which a source's emissions exceed its baseline. For this reason, the credits are referred to in the model rule and by many states as Discrete Emissions Reductions (DERs). Some states also have markets for credits generated by permanent emissions reductions such as plant closures. Usually known as Emissions Reduction Credits (ERCs), these are used to offset the emissions of new sources within an airshed. Differences between these two types of open market credits are obscured by the fact that their names vary between states. Besides the term DERs, the tradable commodities in discrete credit markets are also known, confusingly, as Mass ERCs or Verified Emissions Reductions (VERs). Markets for discrete credits are exclusively the subject of this case study.

   Credits are denominated in mass units of either NO_X or VOCs. Since most baseline regulations are specified in rate terms, the credits must be converted to mass units by multiplying the rate of emissions savings by its duration. In some states such as New Hampshire, one credit is equal to one ton of emissions, whereas in others like New Jersey one credit is equal to 100 pounds, or one twentieth of a ton.

   Credit trading and use are subject to a variety of restrictions. Though the model rule allows for unlimited banking, several states have restricted the length of time that a credit may be saved once it has been generated. Michigan cancels its credits five years after they have been generated. Credits in New Jersey may only be used in the year after they were generated.

10% of any credits used for compliance must be retired as a "contribution to clean air". This is to say that the amount of credits required to demonstrate compliance equals the source's calculated need divided by 0.9. This requirement also may have been included to compensate for discrepancies in calculating emissions reductions. Some states discount credits in additional ways. Massachusetts, for example, discounts credits by a "compliance assurance multiplier" according to the perceived quality of the reductions. Massachusetts also requires users to hold 5% more credits than are needed for compliance.

According to the model rule, users are to bear all liability for the quality of credits. They are required to demonstrate that the credits they intend to use were not previously used and were generated in accordance with state regulations, and they face the risk of penalties for program violations. Some states disregarded the model rule on this point and opted for shared liability.

Because ground-level ozone problems are most acute during the summer months, administrators distinguish between credits according to whether or not they were generated during the "ozone season". Most states allow credits generated during the ozone season to be used for non-season compliance, but not the reverse. The result has been an effective separation between ozone season and non-ozone credit markets. The dates of the ozone season vary from state to state.

Since NO_X and VOCs have local impacts, the direction of trading affects where the program's environmental harms and benefits are realized. Allowing sources to acquire credits from other states, while perhaps cost effective, would work against the importing states' efforts to comply with NAAQS. So interstate trading is generally prohibited or allowed only on a case-by-case basis.

Interpollutant swapping, or substituting credits denominated in one pollutant for compliance with a baseline of another pollutant, is not generally permitted. New Hampshire allows NO_X reductions to be used for VOC compliance, though not the reverse.

3. *Participation.* The program is often called voluntary because no one is required to trade. Any source may elect to meet its baseline emissions requirement without generating or using credits. For some sources, even the adoption of a baseline emissions level is voluntary. However, it is important to note that many sources face what are, in effect, mandatory baselines stipulated in existing command-and-control regulations.

   Market rules state that credits can be generated or used by any source that can establish a measurable baseline in accordance with state procedures. This includes stationary (e.g. utilities, large industrial facilities), mobile (e.g. automobiles), and area sources (e.g. small stationary sources not otherwise included in emissions inventories). In practice, only large stationary sources have shown significant interest in the program.

   DER markets exist in some form in Texas, Connecticut, New Hampshire, New Jersey, Michigan, and Massachusetts. Any person may trade credits generated in the program.

4. *Allocation.* Credits in a baseline-and-credit system exist only if they are created in the course of the program's operation by certified emissions reductions. As such, there is no need for an allocation of credits before the program begins. To the extent that each source's baseline determines its initial level of legally authorized emissions, the process of establishing baselines may be considered analogous to allocation in a closed trading system. For many sources, pre-existing command-and-control regulations will have established this baseline already. Others would have to establish a baseline by documenting their usual emissions levels with a state agency.

5. *Monitoring.* There is no standard emissions monitoring procedure since the program aims to accommodate a variety of sources. The monitoring procedure does not change for sources subject to command-and-control emissions restrictions. Other sources may use whatever practicable procedures exist. Presumably buyers' assumption of full credit liability motivates them to distinguish between credits of different monitoring qualities.

   Credit registries are operated by individual states.[21] These typically

[21] See, for example, New Jersey's <www.omet.com> or Michigan's <www.deq.state.mi.us/aqd/eval/e_trade/registry.html>.

include general information about the program, downloadable forms, and credit availability information including bids, offers, and prices of past transactions. Some states charge users fees for use of the registry.

6. *Certification and Verification.* Sources do not need prior regulatory approval to generate credits. However, the model rule stipulates that generating sources must notify the regulatory authority of their reduction activities within certain time limits. The notification shall be made public, and shall contain, among other things, information regarding: the name and address of the person who owns the generating source, the generation period, the unique identifying serial numbers assigned by the state regulatory authority, a brief description of the generating activity, the amount of credits generated (distinguishing between ozone-season and non-season), the protocols that were used to calculate and document the credits, the source's emissions baseline, and a signed statement that the credits were generated in accordance with required procedures.

   Sources wishing to use credits for compliance must submit a notice of intention to use 30 days prior to the compliance period for which use is intended. During the intervening 30 days, the state may review the proposal to certify its adherence to required procedures. The notice must contain, among other things, information regarding: the name and location of the user, the name of the emissions unit, the permit number, the applicable pollutant, the applicable state and federal requirements that the credits will be used to comply with, the intended use period, the emission quantification protocols that will be used to document the amount of credits needed to demonstrate compliance. The notice must also be accompanied by a copy of the generation notice (described in the previous paragraph).

   Within 90 days after the end of the use period or one year after the beginning of the use period, a source using credits must provide notification of that use. The notice shall contain, among other things, information regarding: the name and location of the source owner, the dates on which the credits were acquired, the amount of credits used and the associated serial number, the use period, the state and federal requirements the credits were used to comply with, and the emissions quantification protocols that were used to calculate the amount of credits required to demonstrate compliance. The notice must be accompanied by a statement that due diligence was made to verify that the credits were not previously used or generated as a result of prohibited activities, a statement that the credits were not used in manner prohibited by law, and a copy of the relevant notice of generation.

Some states specify that the verification of baselines and emissions for official notices must be carried out by approved entities. New Jersey, for example, allows verification only by licensed professional engineers or certified public accountants. Other states such as Michigan allow for self-verification. Typically states require that credits be verified only once in the course of their existence, whether upon generation, while banked, or before use.

Some states grant credit for verifiable reductions achieved before the program's initiation. Connecticut grants credits for emissions reductions as early as 1993. Michigan grants credits as for reductions as early as 1991, but assess a 50% discount to their value. These must be verified according to the same procedures as present reductions.

7. *Enforcement.* Individual states are responsible for reviewing and confirming compliance of individual sources. Violations for use of invalid credits, credit shortfalls, inadequate record keeping, and false certification are subject to penalties under the (federal) Clean Air Act of up to $25,000 per day per violation. Leniency is usually provided in cases involving good faith compliance efforts.

8. *Supplementarity.* As indicated above (2. *Unit of Exchange*), trading is geographically restricted to prevent exacerbation of local pollutant impacts.

9. *Method of Exchange.* States do not regulate the method by which credits are traded.

10. *Timescale.* These markets are intended to operate independently, so their implementation was not contingent on the development of other programs. The model rule was issued by the EPA in 1995, and participating states have subsequently passed implementing regulations based in whole or part on the model rule.

    All sources have been eligible for participation since the program's initiation.

(4) Southern California Regional Clean Air Incentives Market (RECLAIM)

Despite considerable improvements in air quality over the past thirty years, the Los Angeles area of southern California remains out of compliance with federal air quality standards. The Los Angeles Basin is classified as a "severe non-attainment" area for ground-level ozone according to the National Ambient Air Quality Standards (NAAQS) of the 1990 Clean Air Act Amendments (CAA). In the early 1990s the regional regulatory authority for air quality, known as the South Coast Air Quality Management District (SCAQMD), recognized that more stringent emissions restrictions would be required to achieve additional air quality improvements. But use of traditional command-and-control regulations would have been prohibitively expensive. In order to reduce compliance costs, SCAQMD chose instead to achieve tighter emissions restrictions with through an emissions trading program known as the Regional Clean Air Incentives Market (RECLAIM).

Originally the program was meant to regulate emissions of sulfur dioxide (SO_2), oxides of nitrogen (NO_X), and volatile organic compounds (VOCs). Disagreements amongst stakeholders over monitoring procedures and initial permit allocations eventually caused SCAQMD to abandon proposed VOC regulations. Despite similar disagreements over program design issues for the other two pollutants, agreement was eventually reached. In 1994, programs for NO_X and SO_2 were implemented. The programs are unique for their small geographic size, and because they covers a broader range of sources than most cap-and-trade programs. Most stationary sources emitting at least four tons of either pollutant per year are required to participate.

Early in the program, generous permit allocations allowed emissions to exceed previous levels, and trading was less vigorous than expected. More recently, lower allocations have resulted in emissions reductions, and trade volume has shown some growth. Throughout the program, permit prices have been below expectations, resulting in cost savings for affected sources.

■ Description

Design

1. *Market Structure.* Aggregate emissions for sources affected by the program are capped, and the level of the cap declines annually until the program's target is achieved in 2003. The cap remains constant beyond 2003. The eventual target, contained in the region's Air Quality Management Plan (AQMP), aims for aggregate NO_X reductions of 75% and SO_2 reductions of 60% by 2003 relative to the base year of 1994 (for a more detailed explanation of the allocation process, see 4. *Allocation* below).

   Affected sources are divided into a coastal and an inland zone. Trading amongst sources within each zone is unrestricted, but inter-zone trading is restricted to prevent exacerbation of transported ozone impacts. Emissions from the coastal zone cause impacts both locally and in the inland zone where they are transported by wind. So inland zone sources are prohibited from selling allowances to the coastal zone. Coastal zone sources may sell to anyone.

   At the end of each compliance year, sources must demonstrate that they hold at least enough allowances to account for their actual annual emissions. During a subsequent 60 day reconciliation period, sources may continue trading the previous year's allowances to achieve compliance. Sources whose emissions exceed their allowance holdings at the end of the end the reconciliation period are out of compliance and subject to enforcement penalties (see 7. *Enforcement*). Unused allowances expire at the end of their vintage year (see 2. *Unit of Exchange*).

   The markets for SO_2 allowances and NOx allowances are separate, but are administered by the same regulatory body.

2. *Unit of Exchange.* Allowances for both pollutant markets are known as RECLAIM Trading Credits (RTCs). Each confers the right to emit one pound of either SO_2 or NO_X in a particular year. As indicated above (1. *Market Structure*), trading is restricted geographically by zone.

   Allowances are only valid during their vintage year, after

which they expire. Preliminary economic projections suggested that excessive volatility would occur during each reconciliation period if all allowances expired at the same time. So the trading authority divided affected sources into two groups whose allowances expire at different times during the year. Half the sources receive allowances valid for a fiscal year, and half for a calendar year, creating a six-month overlap. Trading between the two groups is unrestricted.

3. *Participation.* Originally, the program was intended to cover about 2,000 sources of VOCs, 700 sources of NO_X, and 50 sources of SO_2. VOC sources were eventually exempted altogether when plans for the VOC market were abandoned. Whole categories of NO_X and SO_2 sources such as landfills and sewage treatment plants were exempted. Other sources were exempted after reviews of their emissions data showed that they did not exceed the minimum threshold for mandatory participation of four tons per year. When the NO_X and SO_2 markets began in 1994, they affected 370 and 40 sources, respectively. These sources accounted for about 65% NO_X and 85% of SO_2 from stationary sources. They include refineries, oil production fields, military bases, electric power, aerospace, chemical manufacturing, glass, steel, cement, and food plants. Additional sources may voluntarily join to the program.

   Anyone may trade RTCs.

4. *Allocation.* Each affected source receives an allocation of allowances on a no-cost basis. The quantity of each allocation is based on a complicated formula that resulted from lengthy negotiations amongst stakeholders. Originally the intention was to grant each source a quantity of allowances equal to its 1994 emissions, and to reduce the allocation annually by a uniform amount that would lead to the program's overall target reductions by 2003. However, California's economy had been in recession from 1991 to 1993, and emissions had fallen as a result. Environmentalists opposed an allocation based on 1994 as a base year

since it would permit more emissions than had been produced in the previous three years. As a compromise, program designers set each source's allocation according to the level of its highest year of emissions during the years 1989 to 1992. This allocation also exceeded recessionary emissions levels, and resulted in an early lack of support from the environmental community.

The final schedule for reducing the cap also resulted from a compromise amongst stakeholders. Some affected sources argued reducing allocations by a uniform annual rate would disadvantage owners of recently installed capital equipment. Whereas the retirement of some very old equipment could be accelerated slightly to accommodate emission reduction obligations, more recently purchased equipment could not be replaced without considerable economic loss. SCAQMD acknowledged this argument, and agreed to conduct an extensive assessment of each regulated facility's equipment. Besides involving many pieces of equipment, the enormity of the assessment was compounded by three exchanges of information between the assessors and equipment owners

TABLE 4.1 Sources Per Annual Allocation-Reduction Rate

Annual Reduction Rates	Number of NO_x Facilities	Number of SO_2 Facilities
2%	93	10
2 to <6%	70	16
6 to <10%	89	3
10 to <16%	118	11
Total	370	40

Source: Lents and Leyden (1996)

in which clarifications and appeals were offered. On the basis of the assessment, annual reduction levels were assigned for each facility through the year 2000. The number of facilities in each reduction-rate category is shown in Table 4.1. After 2000, all facilities' allowance allocations are reduced at a uniform level until the program's ultimate target is reached in 2003.

To initiate the market in 1994, the SCAQMD auctioned 114,676 NO_X allowances and 9,400 SO_2 allowances.[23]

5. *Monitoring.* The rules for monitoring emissions were a source of considerable disagreement between affected sources and administrators. In order to ensure the integrity of the proposed VOC program, all evaporative emissions were to be monitored. This would have required sources to keep track not only of VOC use, but also every instance in which a container holding VOCs was opened. Monitoring accuracy for these materials was not considered accurate enough to assure the integrity of an emissions trading program. So for this and other reasons, the VOC market proposal was abandoned.

   Monitoring provisions for SO_2 and NO_X emissions were also controversial. Monitoring procedures differ according to each source's size. "Major" sources emitting 10 tons or more per year were required to install Continuous Emissions Monitors (CEMs) that measure emissions every minute. Every 15 minutes a Remote Terminal Unit (RTU) calculates an average emissions rate based on these data, and sends this average directly to a computer at SCAQMD. There, an artificial intelligence system analyzes the data and tracks each boiler for compliance. The computer notifies an administrator when the data indicate a potential problem. While this system provides reliable, accurate data, the cost to affected sources ranges from $100,000 to $150,000 for installation of each CEM and RTU. Affected sources initially complained that these charges were excessive. Later they dropped their complaints after economic analyses projected that overall compliance savings would outweigh additional monitoring costs (see 1. *Cost Effectiveness* below).

   "Large" units whose emissions are below the 10 ton annual emissions threshold for installation of CEMs must install Continuous Process Monitoring Systems (CPMs) that measure temperature, oxygen, and fuels use in order to calculate emissions for monthly

[23] Ibid.

reports. Data are submitted monthly. The smallest units were allowed to calculate their emissions on the basis of fuel use multiplied by an emission factor for each type of equipment. Their data must be submitted quarterly. Each site is also visited by inspectors three times per year.

On a quarterly basis every source is required to submit a compliance report, in which emissions data are checked for accuracy by the SCAQMD. Sources are allowed 30 days at the end of the first three quarters to correct any errors. Sources are allowed 60 days at the end of the fourth quarter for the same purpose, and to continue trading the previous year's allowances to ensure their compliance.

SCAQMD maintains a tracking system to record allowance holdings and transfers. Traders are required to register all transfers with SCAQMD after they have occurred. Unlike many other trading authorities, SCAQMD requires that transfer reports include price information. Account information and price data are available on an electronic registry, which also posts bid and offer information from prospective buyers and sellers. The trading authority takes no further role in facilitating transactions, most of which are mediated by brokerages.

6. *Certification and Verification.* Since the program involves allowances, certification and verification functions are fulfilled largely by the emissions monitoring provisions described above (5. *Monitoring*). Quarterly reviews of emissions data constitute a form of certification. Additionally, sources can earn RTCs for emissions reductions from mobile sources such as cars, and area sources not already included in the program. Generation of credits by these means would require verification by an accredited entity, and certification by SCAQMD.

7. *Enforcement.* Sources whose emissions exceed their allowance holdings at the end of the 60-day reconciliation period would be subject to penalties that include subtraction of the amount exceeded from the next year's allocations, civil penalties, and for extreme cases, revocation of the source's operating permit.

8. *Supplementarity.* As indicated above (2. *Unit of Exchange*), trading between sources in the coastal and inland zones is restricted to reduce transboundary ozone impacts.

9. *Method of Exchange.* The SCAQMD does not mandate a particular method of exchange. It does facilitate transactions by publishing transfer price data, bids, and offers. But participants may transact by any means they choose.

10. *Timescale.* RECLAIM's initiation at the beginning of 1994 makes it one of the earliest examples of an allowance-based emissions trading system. EPA's well-known SO_2 Allowance program did not begin until 1995. As such, RECLAIM designers had to address issues regarding aggregate reduction levels and initial allowance allocations that had not been encountered before.

    Sources prepared for the program by testing their new emissions monitoring equipment for a year prior to the first compliance period. Monthly reports were submitted to check monitoring accuracy and to streamline data processing procedures.

    As indicated above, the program was originally meant to regulate emissions of three pollutants rather than its current two. Plans for a VOC trading system remain on hold. In the meantime, VOC emissions are regulated by traditional regulations.

POSITIVE RESULTS

The programs outlined above provide practical evidence of the capacity for emissions trading to deliver positive results for a range of constituents. The public's fundamental interest in protecting the environment by reducing emissions has been achieved as least cost. Regulated sources have enjoyed maximum flexibility to choose their means of compliance with environmental regulations. Government administrators have found emissions trading to be politically attractive, efficient, and simple to maintain.

Emissions trading proponents next face the considerable challenge of reproducing these domestic successes on an international scale with a global greenhouse gas trading system. Though this application will be considerably more complex than past programs, the potential cost savings undoubtedly justify the effort.

Bibliography

Anderson, R.C., and A.Q. Lohof (1997) "The United States Experience with Economic Incentives" in *Environmental Pollution Control Policy*, Washington, DC: US Environmental Protection Agency, <http://206.29.48.66/epalib/incent.nsf/$about>.

Australian Greenhouse Office (1999) "National Emissions Trading: Designing the Market", Discussion Paper 4, Canberra: AGO.

Burtraw, D. (1998) "Cost Savings, Market Performance, and Economic Benefits of the U.S. Acid Rain Program", Discussion Paper 98-28-REV, Washington, DC: Resources For the Future.

Bohi, D. and D. Burtraw (1997) "SO_2 Allowance Trading: How Experience and Expectations Measure Up", Discussion Paper 97-24, Washington, DC: Resources for the Future.

Burtraw, D., C. Carlson, M. Cropper, and K. Palmer (1997) "Econometric Estimates of SO_2 Abatement Costs Under Title IV," in *Proceedings of AWMA/Acid Rain Electric Utilities Conference*.

Carbon Trading Market (2000) "Trading Market", <www.carbontrading.com.au/content/html/trading/main.html?onload>.

Doherty, B. (1997) "Selling Air Pollution: The Politics Go In Before the Market Goes On", *Reason*, <www.reasonmag.com/9605/Fe.BRIANemissions.html>.

Edmonds, J., M.J. Scott, J.M. Roop, and C.N MacCracken (1999) "International Emissions Trading & Global Climate Change: Impacts on the Costs of Greenhouse Gas Mitigation", Pew Center on Global Climate Change, Washington, D.C.: Batelle.

Farrell, A., R. Carter, R. Raufer (1999) "The NO_X Budget: market-based control of tropospheric ozone in the northeastern United States", *Resource and Energy Economics*, 21: 103-124.

Financial Times (2000) "Blair Backs Proposal to Set Up Pollution Permit Market", February 29.

Fisher, C., S. Kerr, M. Toman (1998) "Using Emissions Trading to Regulate U.S. Greenhouse Gas Emissions, Part 1 of 2: Basic Policy Design and Implementation Issues", Climate Issue Brief #10, Washington, D.C.: Resources for the Future.

Guerrero, P.F. (1997) "Overview and Issues on Emissions Allowance Trading Programs", Testimony Before the Joint Economic Committee, Congress of the United States, Washington, D.C.: General Accounting Office.

Hahn R.W. and G.L. Hester (1989) "Marketable Permits: Lessons for Theory and Practice," *Ecology Law Quarterly*, 16:361-406.

Harrison, D. Jr. (1999) "Tradable Permits for Air Pollution Control: The US Experience", in *Implementing Domestic Tradable Permits for Environmental Protection*, Paris: Organization for Economic Cooperation and Development.

International Petroleum Exchange, Enviros Aspinwall, Margaree Consultants (1999) "Design of a UK Greenhouse Gas Emissions Trading System".

Johnson, S. and D. Pekelney (1996) "Economic Assessment of the Regional Clean Air Incentives Market: A New Emissions Trading Program for Los Angeles", *Land Economics*, 72:277-97.

Joskow, P.L. and R. Schmalensee (1997) "The Political Economy of Market-Based Environmental Policy: The U.S. Acid Rain Program", Cambridge, MA: Massachusetts Institute of Technology.

Klaasen, G. (1996) *Acid Rain and Environmental Degradation: The Economics of*

Emission Trading. New Horizons in Environmental Economics series, Lyme, NH: Elgar in association with the International Institute for Applied Systems Analysis.

Kosobud, R.F., D.L. Schreder, H.M. Biggs (2000) *Emissions Trading: Environmental Policy's New Approach*, New York: Wiley and Sons.

Larson, D.F, and P. Parks (1999) "Risks, Lessons Learned and Secondary Markets for Greenhouse Gas Reductions", Policy Research Working Paper, World Bank, <www.worldbank.org/html/dec/Publications/Workpapers/wps2000series/wps2090/wps2090.pdf>.

Lents, J. and P. Leyden (1996) "RECLAIM: Los Angeles' New Market-Based Smog Cleanup Program", *Journal of the Air and Waste Management Association*, 46:195-206.

McLean, B.J. (1997) "Evolution of Marketable Permits: The U.S. Experience with Sulfur Dioxide Allowance Trading", *International Journal of Environmental Pollution*, 8: 19-36.

Mullins, F. (1998) "Lessons From Existing Trading Systems for International Greenhouse Gas Emission Trading", Annex I Expert Group on the United Nations Framework Convention on Climate Change, Information Paper, Paris: Organization for Economic Cooperation and Development.

National Science and Technology Council, Committee on Environment and Natural Resources (1998) "National Acid Precipitation Assessment Program Biennial Report to Congress: An Integrated Assessment" <www.nnic.noaa.gov/CENR/NAPAP/ NAPAP_96.htm>.

Natsource (1999) "Maryland Judge Lays Down the NO_X Law", *Airtrends*, 2: 1.

NESCAUM/MARAMA (1996) Northeast States for Coordinated Air Use Management and Mid-Atlantic Regional Air Management Association, "NO_X Budget Model Rule", <envinfo.com/caain/mact/n158df.html>.

New Zealand Ministry for the Environment (1999) "Technical Design Issues for a Domestic Emissions Trading Regime for Greenhouse Gases", Wellington, NZ.

Nussbaum, B. D. (1992) "Phasing Down Lead in Gasoline in the U.S.: Mandates, Incentives, Trading and Banking", in *Climate Change: Designing a Tradable Permit*

System, T. Jones and J. Corfee-Morlot, Paris: Organization for Economic Co-operation and Development Publication.

Organization for Economic Cooperation and Development (1997) "International Greenhouse Gas Emission Trading", Annex I Expert Group on the United Nations Framework Convention on Climate Change, Working Paper No. 9, Paris: OECD.

Organization for Economic Cooperation and Development (1998) "Lessons from Existing Trading System for International Greenhouse Gas Emissions Trading", Annex I Expert Group on the United Nations Framework Convention on Climate Change Information Paper, Paris: OECD.

Organization for Economic Cooperation and Development (1999) "Tradable Permits for Air Pollution Control: The US Experience", *Implementing Domestic Tradable Permits for Environmental Protection*, Paris: OECD.

Ozone Transport Commission (1994) "Memorandum of Understanding regarding Development of a Regional Strategy Concerning the Control of Stationary Source Nitrogen Oxide Emissions", <www.sso.org/otc/att2.HTM>.

Pilot Emissions Reduction Trading Project (1999) *Emission Trading in Canada: The PERT Experience*, < http://www.pert.org/outgoing/downloads/canada_1.PDF>.

Pilot Emissions Reduction Trading Project (2000) "Draft Rules for Emission Reduction Trading in Ontario", website accessed March 17, 2000, < http://www.pert.org/trading_rule.html>.

Prager, M., T Klier, and R. Mattoon (1996) "A Mixed Bag: Assessment of Market Performance and Firm Trading Behavior in the NO_X RECLAIM Program", Federal Reserve Bank of Chicago Working Paper, August.

Reuters (1999) *An Introduction to Derivatives*, London: John Wiley & Sons, Inc.

Rolfe, C., A. Michaelowa, M. Dutschke (1999) "Closing the Gap: A Comparison of Approaches to Encourage Early Greenhouse Gas Emission Reductions", Vancouver: West Coast Environmental Law.

South Coast Air Quality Management District (1999) *Annual RECLAIM Audit Report for the 1997 Compliance Year*, March 12, <ftp://ftp.aqmd.gov/pub/board/ 990347.exe>.

Stavins, R. (1998) "Market Based Environmental Policies", Discussion Paper 98-26, Washington, DC: Resources for the Future.

Tietenberg, T (1997) "Tradable Permits and the Control of Air Pollution in the United States", written for the jubilee edition of the ZEITSCHRIFT FÜRANGEWANDTE UMWELTFORSCHUNG, <www.colby.edu/personal/thtieten/permits.pdf>.

Tietenberg, T. (1999) "Tradable Permit Approaches to Pollution Control: Faustian Bargain or Paradise Regained?" in Kaplowitz, M.D., ed., *Property Rights, Economics, and the Environment*, Stamford, CT: JAI Press Inc.

Tietenberg, T., M. Grubb, A. Michaelowa, B. Swift, and Z. Zhang (1998) "International Rules for Greenhouse Gas Emissions Trading: Defining the Principles, Modalities, Rules, and Guidelines for Verification, Reporting, and Accountability", United Nations Conference on Trade and Development.

United Kingdom Emissions Trading Group (1999) *Outline Proposals for a UK Emissions Trading Scheme*, October 21.

United Nations Framework Convention on Climate Change (2000a) *Proposed Text for Appendix C, Part Four of the Chairman's Note: Resistries*, Submission by Australia, Canada, Iceland, Japan, New Zealand, Norway, Russian Federation, Ukraine, and the United States.

United Nations Framework Convention on Climate Change (2000b) Appendix C - *Rules and Guidelines, for National Registry Systems*, Submission by Portugal on Behalf of the European Community, its Member States and Bulgaria, Czech Republic, Estonia, Latvia, Lithuania, Poland, Romania, Slovakia, and Switzerland.

United States Council of Economic Advisors (2000) *Economic Report of the President*, Washington, DC: US Government Printing Office.

United States Environmental Protection Agency (1995a) *National Air Quality and Emissions Trends Report*, 454-96-008, Washington, D.C.: EPA.

United States Environmental Protection Agency (1995b) "Open Market Trading Rule for Ozone Smog Precursors", United States Federal Register, <www.epa.gov/docs/fedrgstr/EPA-AIR/1995/August/Day-25/pr-950.html>.

United States Environmental Protection Agency (1997a) Program Overview, Acid Rain Program, <www.epa.gov/acidrain/overview.html>.

United States Environmental Protection Agency (1997b) *Allowance Trading System Fact Sheet*, Acid Rain Program, <www.epa.gov/acidrain/allsys.html>.

United States Environmental Protection Agency (1998) *Compliance Report*, <www.epa.gov/acidrain/cmprpt98/cr1998.htm>.

United States Environmental Protection Agency (1998b) *Lead Phasedown Program*, <http://134.67.55.16:7777/aa/programs.nsf/4ce9.../12b6eb55c5295feb8525651c005 06e2f?OpenDocumen>

United States Environmental Protection Agency (1999) *Preliminary Summary Emissions Reports*, Acid Rain Program, <www.epa.gov/acidrain/otc/otcdata.htm>.

United States Environmental Protection Agency (2000) *Recent Trends in US Greenhouse Gas Emissions*, < www.epa.gov/globalwarming/emissions/ national/trends.html#fnote3 >

United States General Accounting Office (1994) "Air Pollution: Allowance Trading Offers and Opportunity to Reduce Emissions at Less Cost", GAO/RCED-95-30.

Van Horn Consulting, Energy Ventures Analysis, Inc., and K.D. White (1993) "Integrated Analysis of Fuel, Technology, and Emission Allowance Markets", EPRI TR-102510, prepared for the Electric Power Research Institute.

Weber, B.W. (1999) "Screen-Based Trading in Futures Markets: Recent Developments and Research Propositions", Proceedings of the 32nd Hawaii International Conference on Systems Sciences.

Weiner, J.B. (1997) "Designing Markets for International Greenhouse Gas Control", Climate Issues Brief #6, Washington, D.C.: Resources for the Future.

CHAPTER 5

Bandwidth Trading—Developing a Market

By Lin S. Franks and Terrence M. Gee

Bandwidth trading is currently one of the hottest discussion topics on the North American continent and is beginning to be a hot topic globally. Of all the emerging markets, the nascent bandwidth trading market touches the most industries and may affect more people than any market before. Moving information involves using bandwidth—both voice and data. The ability to create and process bandwidth is what defines one's ability to work in an information society. The market will ultimately be a global market, however, the immediate interest is intensely North American. This seems reasonable when taking into account that 70% of the Internet is North American-based.

Bandwidth determines the speed at which data can flow through computer and communications systems without interference and is bought and sold in several different interface speeds such as T1, T3, OC3, OC12, and OC48. A Digital Service 3 (DS3), as proposed by Enron Broadband Service (EBS) as the initial product to begin to trade, is a dedicated phone connection supporting data rates of about 43 Megabits per second (Mbps). A DS3, which is also known as a T-3 line, actually consists of 672 individual channels, each of which supports 64 Kilobits per second (Kbps). DS3 lines are used mainly by Internet Service Providers (ISPs) connecting to the Internet backbone and for the backbone itself.

On October 24, 1995, the Federal Networking Council (FNC) unanimously passed a resolution defining the term "Internet". This definition was developed in consultation with the leadership of the Internet and Intellectual Property Rights (IPR) Communities and states that the Internet refers to the global information system that: (1) is logically linked together by a globally unique address space based on the Internet Protocol (IP) or its subsequent extensions/follow-ons; (2) is able to support communications using the Transmission Control Protocol/Internet Protocol (TCP/IP) suite or its subsequent extensions/follow-ons, and/or other IP-compatible protocols; and (3) provides, uses or makes accessible, either publicly or privately, high level services layered on the communications and related infrastructure.

The delivery system for the evolving commodity "bandwidth" is the Internet, the most widely used network in the world. It was initially developed in the 1960s as a data network for linking computers at universities and government agencies around the United States and has grown to include more than 100 million users worldwide today. The Internet is a network of networks.

Events of the past year have caused an unprecedented level of interest in bandwidth trading for a market at the very beginning stages of its development. It is of particular interest right now to US energy companies looking for ways to elevate the perceived value of their stock. The series of events that caused this interest to rise began in the spring of 1999 when Enron Communications Inc., now known as Enron Broadband Services (EBS) announced they were creating a commodity out of bandwidth. Since that time, a few deals were executed between EBS and others in the industry under an Enron proposed "straw man" agreement. Enron proposed an agreement with standard terms and conditions for a single bandwidth product, DS3, as the initial product to trade. In early 2000 stock analysts began to take notice of the developments in the area of bandwidth trading and gave Enron stock a very positive valuation as a result. Enron enjoyed a subsequent stock lift, which evoked the interest of other energy companies who also own communications assets and many who do not.

MARKET PARTICIPANTS

In the communications market, participants fall into three classes of entities: facilities owners, resellers/integrators, and consumers. Facilities owners actually own the physical infrastructures that make it possible to move bandwidth around including the fiber optic plant, the copper plant, the cable television plant, satellites and microwave installation. These are companies like AT&T, MCI Worldcom, and Sprint. Resellers are carriers without facilities. They do not actually own the infrastructure but they own the rights to use other people's infrastructures. They buy bandwidth capacity from others, install some of their own switching equipment and use the capacity for their own economic benefit. These include companies like Touch America, and other second and third tier providers. The bargain long distance rates advertised in the newspapers such as 5 cent a minute is usually from a reseller. The last category is the consumer. Consumers do not own facilities but require bandwidth to deliver services to their own customers or for their own use.

The various ISPs are segmented into groups that have common characteristics. This is largely a reflection of the very hierarchical manner in which the TCP/IP addressing protocols work. It is also a reflection of relative commercial power, of reach, and relative value

Using a pyramid hierarchical structure, at the bottom of the pyramid are the regional ISPs. These providers are focused on offering service in a geographically limited area, often within a single city. They typically offer dial-up services, with ISDN and xDSL access available in some areas. They may offer dedicated line services to limited numbers of customers. These providers compose the largest segment of the market by number. However, in order for them to be able to serve their customer base, they must have a transit agreement with a larger provider. A transit agreement is a contract for service in which the larger provider establishes a dedicated line connection to the smaller provider. The larger provider also agrees to accept packets destined for points outside the smaller provider's network and have them delivered appropriately and conversely deliver any packets addressed to the smaller provider's network.

These larger providers may be mid-tier ISPs or National Service Providers (NSPs). The mid-tier ISPs are best thought of as regional ISPs that expanded to cover a larger region, often a multi-state area, but stop short of having the infrastructure to elevate them to the next level. They too typically must purchase transit agreements from still larger upstream providers, but often have peering agreements with other similar sized providers.

At the top of the pyramid are the "Tier One" backbone providers. These providers are those who do not require transit agreements with any other providers, because they have a large enough network to have allowed them to gather substantial quantities of peering agreements. From a "Tier One" view, the world consists of other "Tier One" providers and their customers.

It is assumed that any customer of a Tier One provider can be reached through that provider. Therefore, "substantial" means that for any legal TCP/IP address, the destination is either to (1) a given Tier One provider, (2) to the customer of that same given Tier One provider, or to a Tier One provider with which peering relationship exists, or (3) to the customer of another such Tier One provider.

BANDWIDTH AS A COMMODITY

Currently when companies buy and sell bandwidth each deal is negotiated as a unique agreement. There is no standardization of terms and conditions to simplify the procedure with the exception of the Enron standard DS3 agreement. Negotiations are counter party to counter party. In a typical transaction the consumer goes to one of the providers, either in the retailer space or in the facility owner's space and contracts for service. Every provider has their own terms and conditions which differ from another provider's terms and conditions. You may even see different terms and conditions from the same provider depending on which piece of the organization you are dealing with. It takes anywhere from days to months to conclude the negotiations for the contract. Once the contract is executed, it may take anywhere from a matter of a few weeks to around six months for the facilities to actually be made available for your use. This day-to-day experience together with risk management and trading skills is valuable but should be complimented with knowledge of emerging markets and the process of commoditization.

For a product to become a "commodity" it must be both consumable and fungible. Bandwidth is consumable, but currently it is not fungible. To be fungible, it must be exchangeable one product for another such as natural gas where one molecule is exchangeable with another natural gas molecule. Bandwidth is bought and sold between city pairs such as Los Angeles to New York or New York to London. One cannot exchange bandwidth between Los Angeles and New York for another city pair or another interface speed on an equal basis. It is therefore not fungible. However, it is possible with standardized contract terms and conditions to vastly improve the cycle time for buying and selling the various products (different city pairs or different interface speeds). Over time, one product may emerge as a benchmark product. All other products may then be priced in relationship to the benchmark product. A liquid market for these differentiated products could be facilitated by utilization of an electronic platform such as has been launched for natural gas, crude oil and electricity, however, it needs a little time to develop before it makes sense to invest in such a platform.

MARKET DEVELOPMENT

The bandwidth trading market development is just beginning. Markets that have undergone the commoditization process before it have completed the process with increasing speed. For example, in North America, the "commoditization" of natural gas took approximately 25 years. The "commoditization" of electricity, which followed natural gas, took only 3 years. It is reasonable to assume then, that the "commoditization" of bandwidth will take even less time...perhaps as little as a year.

The Internet is rapidly becoming the principal communication and marketing tool of world commerce. E-commerce solutions for business frequently involve sophisticated intelligence built into the software in order to simplify the business processes for users and graphics to enhance the experience. This intelligent and graphics intensive software requires large amounts of bandwidth to meet the expectations of today's user who demands instant gratification and a video game presentation. Most industries in the US are either utilizing or plan to utilize the public Internet to deploy mission critical software. Their performance is therefore dependent upon being able to meet the ever-increasing bandwidth demands of their growing client base. Projection of supply and demand indicate that within a couple of years that demand will significantly outstrip available supply even considering the rapid and seemingly continuous build outs and technological advances that promise to increase the bandwidth capabilities of existing networks. This imbalance between supply and demand is the principal driver of the creation of a more efficient market in the buying and selling of bandwidth...or the "commoditization" of bandwidth.

It is important for energy companies with communications assets to begin to become involved with the development of this market. It is equally important for telecommunications companies, companies with a primary source of income is derived from e-commerce, and energy companies without communications assets to participate in how this market emerges. They will all have a role in this market.

Energy companies with communications assets who also have a core competency in risk management and trading are particularly well-positioned to take advantage of this emerging market. Telecommunications companies who elect to participate will gain an additional market in which they will be able to buy and sell for shorter terms and with a number of potential partners unavailable to them in the traditional market. They may require risk management and trading skills that do not already reside within their organizations to compliment their deep comprehension of telecommunications industry fundamentals and it's underlying technology.

TRADING AND RISK MANAGEMENT

While trading and risk management skills are extremely valuable long term, the skill of buying and selling bandwidth day-to-day in the traditional market is critical in the beginnings of this market. Companies with both skill sets enjoy an ability to leverage traditional bandwidth buying and selling expertise short term and trading and risk management skills longer term. These trading and risk management skills are best applied once the market has some standardization, liquidity, price volatility, and a forward curve.

Most risk management and trading talent are more experienced working in developed markets than in emerging markets. They expect the existence of forward price curves and publicly available pricing information. They also expect the negotiation of a deal to be more about price than about any other factor. These things do not yet exist in this market. Risk management and trading expertise is essential for a market with price volatility. Currently there is no price volatility in this market. There is an exponential decline in price of approximately 40% per year. Negotiations for bandwidth are as much about Quality of Service and several other factors as they are about price. Delivery risk is probably more significant than price risk. Therefore a deep fundamental understanding of the way bandwidth is bought and sold traditionally may be more valuable than risk management and trading expertise in the short term.

Once this market evolves, companies that derive significant revenue from e-commerce who enjoy an increasing client base will be able to secure additional bandwidth to meet short term needs without the magnitude of capital investment required in the traditional market. The cycle time for negotiating and provisioning will also be dramatically shorter than the industry currently experiences. Provisioning is the process of identifying, assigning and joining all of the transmission paths that comprise a circuit. Currently this process can take from days to months to complete. The bandwidth trading market has the potential for promoting nearly instantaneous provisioning.

Some of these consumer companies may elect to purchase a flexible plan from their ISPs; however, some may elect to secure this product for themselves. Even if they secure a flexible service from their ISP rather than purchasing from the market themselves, the existence of this market will make it possible for the service to exist at a reasonable price. Companies, for whom the acquisition of adequate bandwidth is critical to serving their customers or users, may use this incremental market to acquire reliable bandwidth for shorter periods of time (such as for peak needs) without committing to traditional long-term agreements. It will also provide a mechanism for selling excess bandwidth for short periods and for acquiring bandwidth in geographic areas where they currently have no facilities.

BANDWIDTH AND ENERGY

Perhaps one of the most important reasons for energy companies and others to enter this market is that electricity is the fuel for the Internet. Today, the Internet consumes approximately 8% of US electricity. Ten years ago that % was zero. In the foreseeable future it is not unreasonable to expect the Internet to consume from 30-50% of US electricity. 1 Some industry experts believe that this number is conservative. Even if it is overstated, it is still a significant potential impact. If it is conservative, then it truly changes the risk profile of all companies with a tie to the Internet.

Electricity is generated from fuels such as oil and natural gas. So from a value at risk perspective, some companies will not be hedged unless they hedge all commodities in the value chain: Oil, natural gas, electricity, and bandwidth. Others will need to hedge only electricity and bandwidth. Currently power outages are concurrent with outages on the Internet. With power prices spiking to unbelievable numbers during summer months and the speculation that the Internet will consume more and more power, then the need to hedge also increases.

MARKET ENTRY STRATEGIES

Whether an energy company or a telecommunications company, success in this market will require sound business planning and execution. Three fundamental elements must be in place to be positioned for success: (1) a Market Entry Strategy, (2) Business Architecture, and (3) a Performance Support Program

This market will not be for the faint of heart or the ill prepared. Success will require careful consideration of the appropriate market entry strategy. Organizations must ask the tough questions, such as "what's my appetite for risk"?, "how well-positioned are my assets?", "can I strike the right alliances to make up for any short-coming?". The answers should provide guidance for determining several potential "participation" scenarios that might evolve. Each scenario needs to then be thoroughly "stress-tested" to ensure that it's pursuit will truly result in the desired financial payoff, whether measured in terms of net income of stock price appreciation.

Once a company has defined the appropriate strategy, the hard work is just beginning. It's imperative to create the right business architecture. The business architecture defines how an organization's people, executing well-defined work processes and enabled by the right technology, can achieve the targeted value.

Because the market is new and will be evolving for some time, creating the appropriate performance support program is also critical. Well-defined strategies, clear performance objectives, the latest technology won't translate into success if the workforce doesn't have the resources to ensure that they obtain the right skills and stay current as the market requirements evolve.

CONCLUSIONS

Active participation by all industry segments is required for this market to develop in a manner that supports the industry. As demand for bandwidth increase so will volatility. As volatility grows companies will require a benchmark contract with which to hedge. Enron proposed an initial product to trade: the DS3 between Los Angeles and New York. This may or may not be the product that evolves into the benchmark product. Industry task forces and self-governance organizations have begun to emerge to address this and related issues. Active involvement by all industry segments may result in a different product as the benchmark. For example, that benchmark could be an OC48 and could be tied to a set number of miles rather than to a specific city pair. This benchmark product design would lend itself to simple scaling to fit all city pairs (See Box).

In summary, the bandwidth trading market is developing at this time because supply and demand are significantly out of balance now and in the foreseeable future. In the June 1999 report published by the Greening Earth Society entitled "The Internet Begins with Coal", Mark P. Mills states that the electricity appetite of the equipment on the Internet has grown from nothing ten years ago to 8% of the total US Electricity consumption today. He further states that at some point in the foreseeable future it is not unreasonable to assume that the Internet will consume 30-50% of the U.S. energy supply.

The skills needed to take advantage of this emerging market are: (1) Risk Management and Trading, (2) Options, and (3) Industry Fundamentals. The market is of interest to and will provide value to companies who (1) have both energy risk management and trading skills together with communications assets, (2) are asset based communications service providers, (3) deploy mission critical software on the public Internet, or (4) provide electricity, the fuel for the Internet. This market is at its initial stages of development, but is expected to develop more rapidly than the electricity market, perhaps as quickly as one year. Because of this expected rapid development it is essential for affected entities to participate actively in industry endeavors around standardization and trading.

AN EXAMPLE APPLICATION OF A MILEAGE BASED BENCHMARK

Bandwidth is priced in cents per DSL (digital standard line) mile. A DS0 or digital standard "0" is equal to one voice channel. Airline miles are used for pricing between city pairs.

There are:

- 672 DS0 per DS3.
- 24 DS0 per DS1
- 28 DS1 per DS3
- 48 DS3 per OC48

If the price is 1.5 cents per DSL mile then the price for one mile of OC48 is 0.015 X 672 X 48 or $483.84. Similarly, if the benchmark contract is for 100 miles, then the contract value would be $48,384. The airline miles between New York and Los Angeles are 2800 so the price of that city pair is $1,354,752. The airline miles between Houston and London are 4842, therefore the price based upon the benchmark contract is $2,342,753.30.

In this example, the benchmark contract is for 100 miles rather than one mile. This is in anticipation of the benchmark contract trading as a commodity trades...with both speculators and hedgers participating in the market. For the market to attract a large number of players the contract value needs to be small enough so that smaller speculators can afford to trade but large enough that transaction costs are not a burden. If the benchmark were only for one mile then it would require 280 contracts to hedge a position in OC48 between New York and Los Angeles. At 100 miles per contract, only 28 are required.

The benchmark may also evolve as the industry evolves. One proposition is that the benchmark be a lambda contract or a light wave contract. A lambda is equal to an OC48 without the electronics. It might be priced at 0.25 cents per DSL mile. At a contract quantity of 100 miles it would have a value of $8,064. A light wave is fungible.

[1] In the June 1999 report published by the Greening Earth Society entitled "The Internet Begins with Coal", Mark P. Mills states that the electricity appetite of the equipment on the Internet has grown from nothing ten years ago to 8% of the total U.S. Electricity consumption today. He further states that at some point in the foreseeable future it is not unreasonable to assume that the Internet will consume 30-50% of the U.S. energy supply.

CHAPTER 6

Bandwidth Trading—The New Commodity Gold Rush

By Michael Moore

The commercial reality of bandwidth trading has evolved through a combination of market events and drivers over the past year. The events that have catalyzed the markets emergence primarily relate to the exponential increase in Internet usage. Notable drivers include America on Line (AOL)—the world's largest Internet Service Provider (ISP)—exceeding one million customers, followed by other ISP's beginning to run out of capacity to service customer demand. A direct result of increased demand for Internet usage was the increased demand for additional phone lines, leading to congestion at local telecommunication companies. And, as the demand to link into the Internet increased, so did the need for speed, as customers demanded faster and more efficient Internet access. As more and more businesses become web-enabled, providing Internet access for all employees, additional capacity for Internet connectivity is, in itself, insufficient. The commercial reality is that these businesses require fast, continuous and reliable connectivity to the Internet, which has hastened the evolution of bandwidth trading.

In tandem with the telecommunication demands of business have been the advancements in technology, which are facilitating the evolution of bandwidth trading. This includes more efficient routers, faster computers with more memory to facilitate the more robust browsers, and faster modems. As Internet connectivity increased more fiber was laid, more technology was deployed, more commercial users surfaced, and the demand for bandwidth took off. This in turn led to the creation of a plethora of new applications that demanded still more capacity. The result is that US demand for bandwidth is doubling every 72 days. In December 1999 there were over 110 million online US users with web pages being added at an average rate of 2 million per day by March 2000.

TABLE 6.1

A LOT CAN HAPPEN IN 24 HOURS ON THE WEB A Snapshot of the Web's Size, 1999		
Metric	**1999 Total**	**Estimated Growth Per Day 1998-1999**
World population	5,996,708,634	213,188
Web pages	1,500,000,000	1,917,808
Devices accessing the Web*	221,100,000	196,164
Worldwide Internet users	196,100,000	147,671
Hosts	72,398,092	79,913
Domain names	8,100,000	12,981
Unique Web sites	3,649,000	4,422

This has put a tremendous amount of pressure on companies like to Lucent, Nortel, Ericsson and Cisco to constantly come up more innovative ways to stay up with this demand. According to Lucent the current equipment cycling or obsolescence rate for routers, switches, servers and modems is running at about 24-30 months. This allows new technology to get deployed rapidly, so demand is being met...but look out for the day that the demand exceeds capacity in a market that can price itself in real-time. The same push on hardware is evident in fiber optics and the related optical switches that are being introduced. Today, using Dense Wave Division Multiplexing (DWDM), hundreds of separate signals can be sent over one fiber. This is accomplished by combining optical signals of different wavelengths (or colors of light) onto a single fiber. DWDM has effectively been used to increase the capacity of the existing fiber by a factor of 16 with improvements of 80 times or more possible. Thus, bandwidth is increased through the existing infrastructure of fiber optic cable. It is only one of many technological developments underway to put the bandwidth infrastructure in place for market-based trading.

WHAT IS BROADBAND?

The Broadband revolution is attracting many new market participants, but it is helpful to know what broadband actually is. Broadband is a general term that refers to high-speed data communications at which users can access the Internet, and is benefiting from the increased worldwide deregulation of the telecommunications industry, which is unleashing rapid technology transference and a building out of new capacity.

MARKET DYNAMICS

The market dynamics of bandwidth allow parallels to be drawn from other commodity markets. With so much demand driving so much new technology one would think that this is unlike any other commodity market around. Almost, except the demand comes from humans and their demand patterns for anything that they consume is fairly consistent. When we got to bed, we lighten up our overall power consumption. Similar consumption patterns are evident during holiday periods, inclement weather and a host of other 'normal' occurrences. This profile has been know for sometime, but you would have to be in a market like electricity to appreciate the dynamics to drive it and some of the "unique" events that surface in a real time commodity market.

Consider the following charts for electricity and telecommunication volumes.

FIGURE 6.2

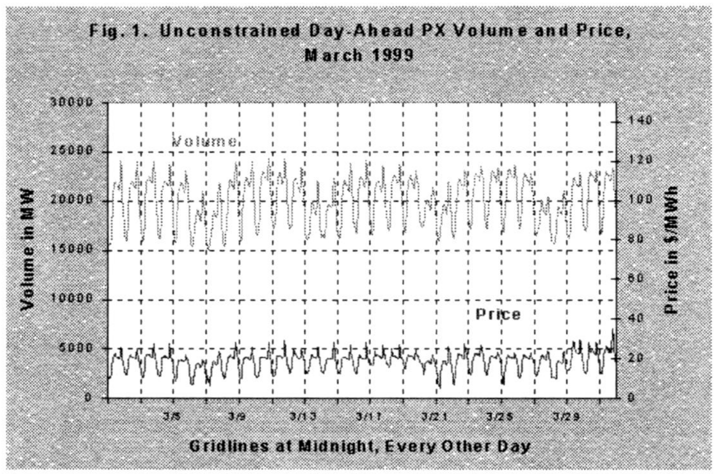

FIGURE 6.3

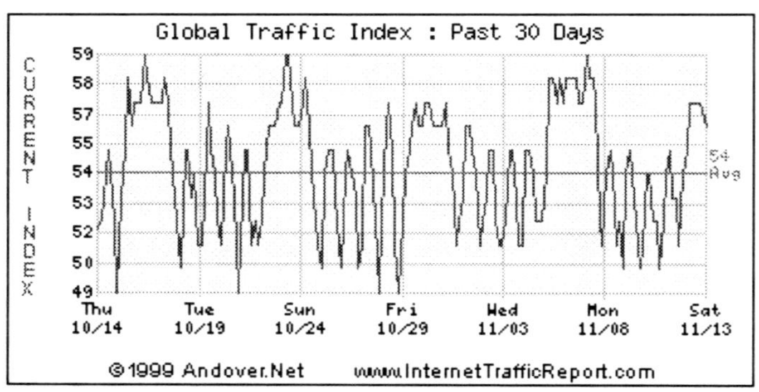

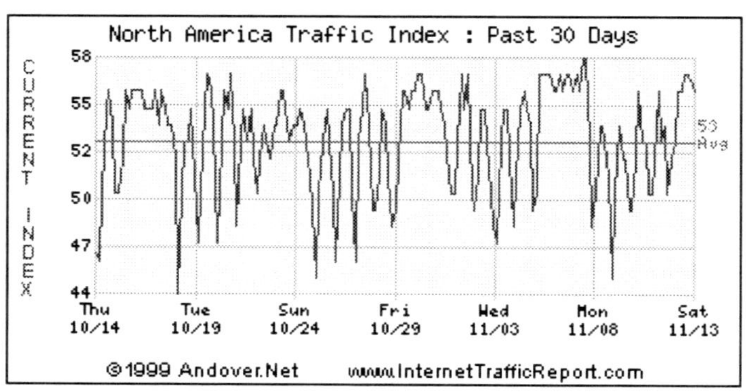

The influence of regional events such as sunrise and sunset, particular working habits, the type and speed of Internet access, and the technology that facilitates Internet access all determine telecommunication demand patterns. Demand patterns are also influenced by issues such as whether or not a geographic area is economically conducive to be online, and local taxation issues, which have an impact on consumer traffic patterns between New York City and New Jersey due to the sales tax differences. Demand patterns are dependent on how these issues are handled. The end result is a variety in demand, which fluctuates over a 24-hour period. In power the demand fluctuation determines the hourly prices that are paid and sold in each region, which creates tremendous trading and arbitrage opportunities for market participants.

TABLE 6.4

SONNET Signal Hierarchy		
Level	**Line Rate**	**DS3 Channels**
OC-1	51.84 Mbps	1
OC-3	155.52 Mbps	3
OC-9	466.56 Mbps	9
OC-12	622.08 Mbps	12
OC-24	1.244 Gbps	24
OC-48	2.488 Gbps	48
OC-96	4.953 Gbps	96
OC-192	9.953 Gbps	192

Digital Signal Designation	T-Carrier Designation	Transmission Rate	#of Voice Circuits
DS-0	–	64 Kbps	1
DS-1	T1	1.544 Mbps	24
DS-2	T2	6.312 Mbps	96
DS-3	T3	44.736 Mbps	672
DS-4	T4	274.176 Mbps	4032

SHAPING THE MARKET

So what are the current events that are building the momentum in bandwidth trading? The primary catalyst is a new and previously unnoticed player entering the telecommunications market—energy companies. With seasoned trading desks and the mentality to aggressively take on a new market companies such as Enron, Williams, Koch, Columbia Energy Communications, Duke Communications, Southern Company, El Paso and

Dynegy are all addressing the opportunities in the communications market. All these companies have something in common: they already have a vertically stacked trading and risk management infrastructure and they will take on one of several strategies to develop the market. The primary strategies involve either the acquisition of a telecommunications company, or the monetization of current communications assets—existing networks, customer bases or energy pipeline companies using their right-of-ways for fiber optic routes. No matter which route is taken it puts the energy companies into the existing telecommunications company's back yards. Yet, because of this potential threat of new players with new technology and seasoned trading experience, you are seeing them move into this trading area very fast as well.

On March 13, 2000, H. Russell Frisby, president of Comptel, announced that it had set up an ad hoc task force to begin the work needed to standardize a bilateral master agreement, identify benchmarks to create a standardized product, and begin the formation of a Bandwidth Trade Organization (BTO). The concept is that the BTO would be comprised of telecommunications carriers, large end-user, traders, pooling point operators and administrators. This organization would assist in the development of standardized contract terms and conditions for the trading, measurement and provisioning of bandwidth. Comptel is the taking the role of power industry's equivalent of WSPP and its early work on a standardized enabling agreement between members that facilitated early trading of the deregulated marketplace. These are the precursors to market development.

Enron announced its intention in May 1999 to transact the first "commodity" type bandwidth trades, using terms and conditions including "Master Agreement", "Firm with Liquidated Damages", performance and credit terms. This was not a "business as usual" proposition for the traditional telecommunication utility and was met with general skepticism. By the first week of December 1999, that first trade had taken place and there were more to follow. Today there are several others that have followed Enron's initiative and numerous transactions have followed. There are also at least three over the counter (OTC) brokers that are using their people and skills to develop liquidity and market awareness similar to the early genesis of the power and natural gas markets. This development process will ultimately lead to these products being transacted on electronic trading platforms.

What caused this event? As the carrier market grew in participation,

FIGURE 6.5

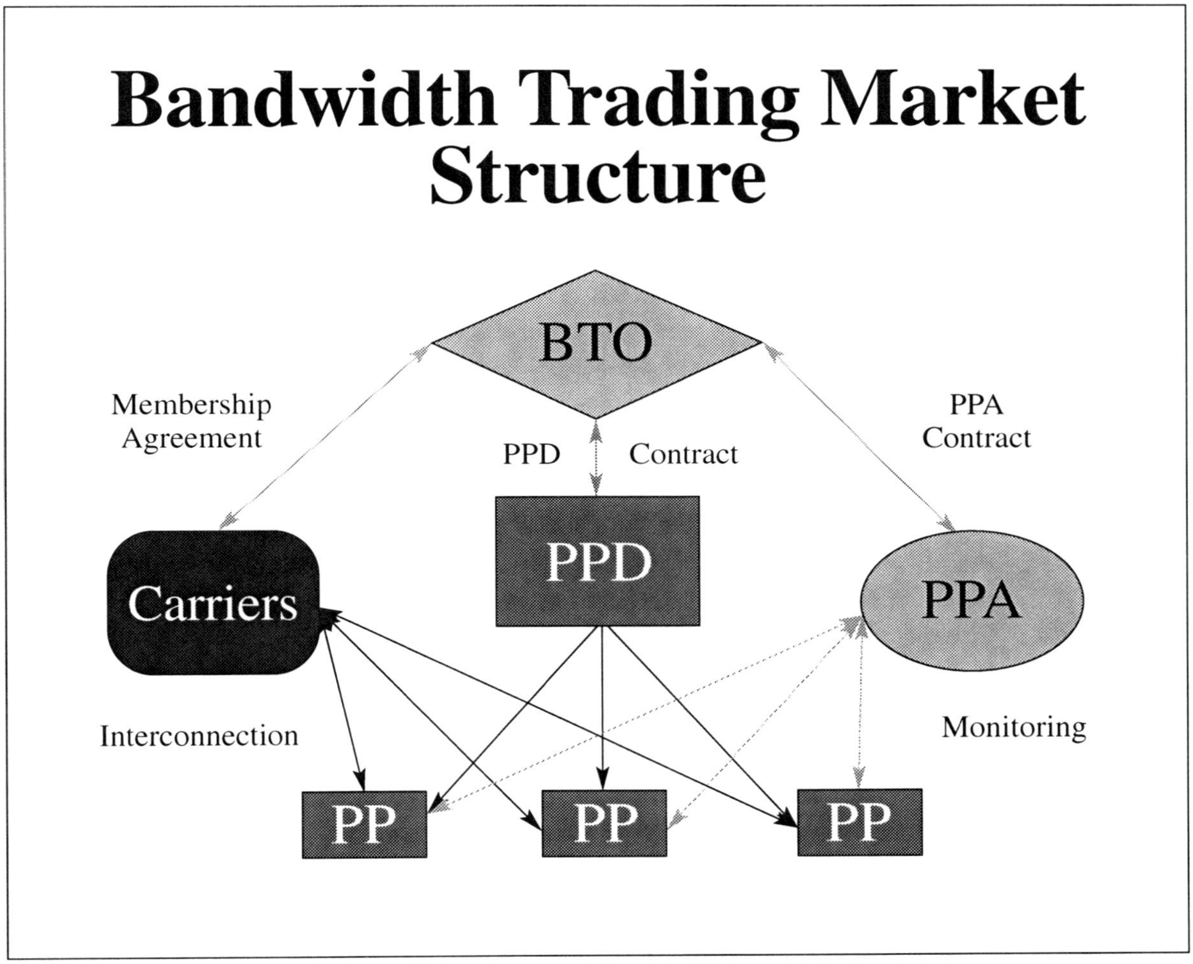

so did the technology that created a natural arbitrage between old and new equipment in the field. While this was happening, more and more companies began using the web for their actual core business activities and wanted more flexible terms with their access providers, who in term have started to demand it from the carriers. Companies like Universal Access and Exodus sprang up to provide more flexible services to the commercial end users. With web traffic growing so rapidly, no one has been able to predict where or when the next "Toys-R-Us" event will take place, so they all want the ability to respond. In November 1999, Toys-R-

Us launched one of their biggest pre-Christmas ad campaigns and they were so overwhelmed by the response that their web servers all went down and subsequently took the network down nearby as well. As a result the company completely lost its Christmas 1999 window for sales, and Wall Street responded accordingly.

BANDWIDTH MARKET DEVELOPMENTS

What is the market today and where is it going? There are currently numerous carriers and service providers working on the necessary framework to participate in this new business sector. Houston-based energy brokerage Amerex has been spending high quality time with these potential new customers, helping them to source information and to inform them just what a commodity market is and what to expect. There are several other OTC brokers doing likewise such as Sakura Delsher and Prebon-Yamane. Other factors include the work being conducted by LighTrade, which is developing itself as the "pooling point" or market hub in most of the major cities in the US, and is looking to expand into the Far East and Central Europe.

It is these events that make a benchmarked fungible product a reality. With this will evolve price visibility and price indices that will allow the formation of a derivatives market similar to the existing market in natural gas and the emerging market in electricity. This evolution will attract non-traditional players into telecommunications, and will allow for a more flexible community of products that the end users can access. The size of this market, according to Shroeder's, it is estimated to be $350 billion, making it larger than power and natural gas.

Another effect of this new market is the changing attitudes and expectations of traditional lenders. To simply walk into a financial institution and provide proof of a forward sale for 10-20 years of firm capacity on a proposed project will not receive the same attention today that was there before. It has become a "catch-22" for both new entrants and established participants. If the analyst community feels confident that this will become a commoditized market like electricity and natural gas, then it will expect the development of hedging and risk management tools that are common in the gas and electric markets. This expectation will create a different attitude towards a company that is resisting the process of commoditization,

FIGURE 6.6

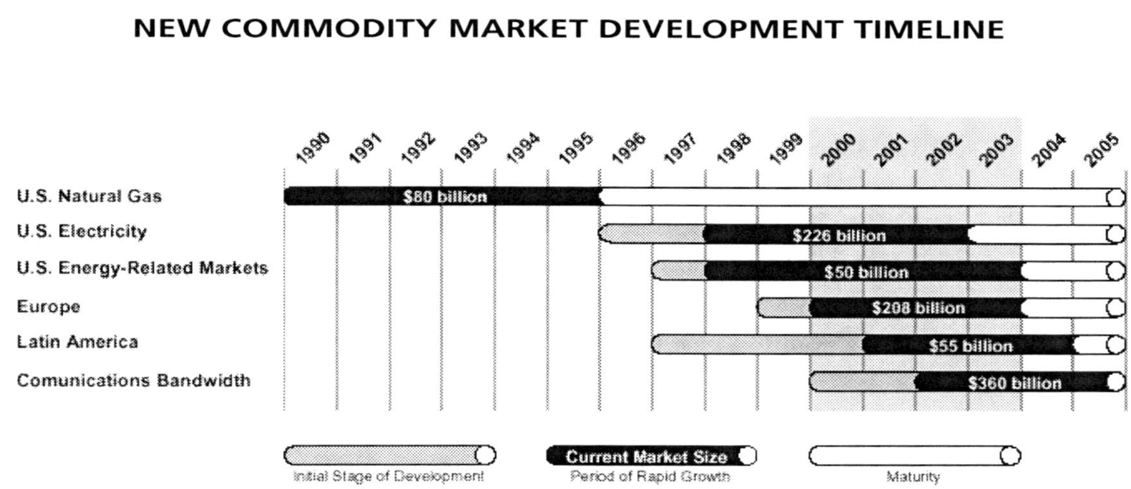

Source: Schroder & Co., Industry sources.

and reward significantly those that embrace the event and move into the forefront.

Enron recently experienced this by the huge gain in share price from mid-December through mid-January, going from a traded value of the high $30's to briefly reaching $74. This has also resulted in a sudden interest at board level from several other companies with similar abilities to openly pursue a strategy towards encouraging the commoditization process of bandwidth by participating in the formation of the BTO. Because of the open retail aspect of this and the rapid deployment of "last mile" solutions, there will be price uncertainty that the experienced commodity participants will feel right at home with.

VOLATILITY

As the pool of companies actively involved in the commodization process increases, the volatility that can prevail in a global real time market will surface. The primary drivers of volatility will be the time zone shifts of demand, where it originates and where that demand will need access. Since most public web pages are in the US (70% of the global market), they reside on US-based servers. As the day progresses the level of European demand access to these web pages falls as day turns to night. Demand then switches to the Far East as demand follows the movement of the sun. Now throw in a couple of major holidays that are not globally shared, or regional targeted advertising or commercial events along with some equipment failures, and you can begin to get some interesting volatility.

FIGURE 6.7

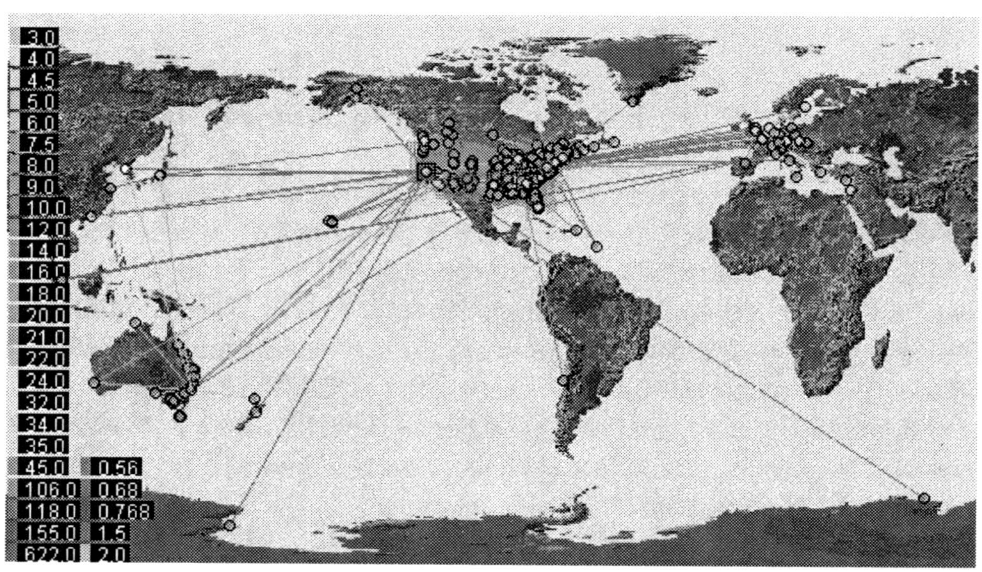

FIGURE 6.8

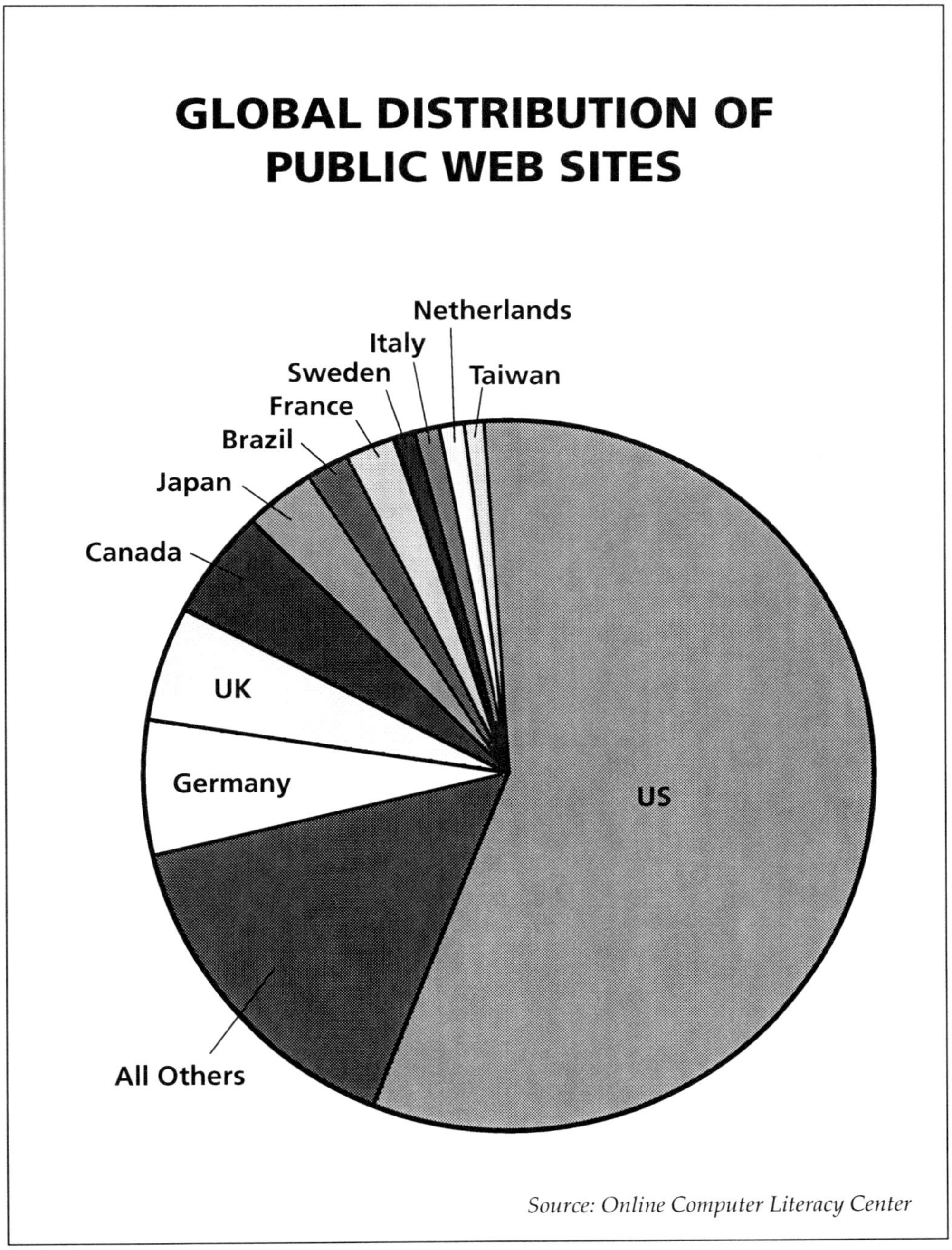
GLOBAL DISTRIBUTION OF PUBLIC WEB SITES
Netherlands
Italy
Sweden
Taiwan
France
Brazil
Japan
Canada
UK
Germany
US
All Others
Source: Online Computer Literacy Center

CHAPTER 7

European Natural Gas: Towards a Competitive Market

By Seana Lanigan

The structure of European natural gas markets, like electricity, is undergoing a process of change. For some this represents a revolutionary process of reform while for others it is an evolutionary process. European Union (EU) legislation is forcing many countries to break up state monopolies and introduce at least a degree of competition into their domestic markets. As a result of legislation the hope is that competition will extend internationally with the ultimate end result of lower consumer prices while ensuring security of supply is maintained.

THE EUROPEAN GAS DIRECTIVE

The stated aim of the European Gas Directive is to establish a competitive European gas market, which will be one of the final steps towards finalizing the end vision of a single European energy market. Common rules are intended to provide a competitive market structure, which will be implemented in detail by the individual EU Member States, in line with the principle of subsidiarity. European energy ministers finally reached unanimous political agreement on the EU's proposals for a competitive gas market following 18 months of negotiations in December 1997. Problems had arisen because of the variety of different market structures in each of the Member states. Formal adoption followed at the Energy Council of 11 May 1998, and Members states have been given just over two years to bring the necessary national legislation into force.

The EU directive sets out rules for the transmission, distribution, supply and storage of natural gas. It covers the organization and functioning of the sector, including liquid natural gas (LNG), market access, system operation, and the criteria and procedures for the granting of authorizations for transmission, distribution, supply and storage.

MARKET OPENING

Market opening will be achieved through a combination of eligibility criteria for certain consumers to have access to competitive supplies, with percentage targets based on total annual gas consumption of Member states. All gas-fired power generators and other final customers (including co-generators) consuming more than 25 million cubic meters of gas per year per site, should be eligible once the directive is implemented.

The initial 25 million cubic meters threshold reduces to 15 million cubic meters per year after five years, and 5 million cubic meters after ten years (i.e. by June 2008). By this means, it is intended that there should be an initial market opening equal to at least 20% of the Member state's total annual gas consumption, increasing to 28% in 2003 and to 33% in 2008. These staged threshold levels mimic the opening of the EU electricity market in 1999 and as such recognize the development of a convergent gas and electric market in Europe.

If the eligibility criteria lead to an initial market opening of more than 30% in a Member state it can be limited to 30% or more, and to 38% and 43% or more in the second and third phases, whilst avoiding discrimination against particular consumers. Analysis of the European Commission's survey of power generators and eligible consumers suggests that the directive will lead to initial EU market opening of nearly 34%, rising to just over 36% and finally to over 42% over the above timetable. Again, as with the European market's experience with electricity liberalization, market opening will exceed the stated minimum thresholds.

NETWORK ACCESS

Member states may choose between a system of negotiated access or regulated access, both operating with objective, transparent, and non-discriminatory criteria. Most will choose negotiated access, under which natural gas undertakings will have to publish their main commercial conditions for the use of the system. This will be required within the first year following implementation of the directive and annually thereafter. The definition of main commercial conditions is open to interpretation by Member states.

TAKE OR PAY CONTRACTS

The existence of long-term take or pay gas contracts was agreed to provide legitimate grounds for refusal of access to networks. It has been said that such contracts safeguard security of gas supply in the EU and provide the market stability necessary to attract high levels of investment. However, the Commission and those in favor of market liberalization were concerned that long-term take or pay contracts should not be used to prevent or delay market opening. A procedure has therefore been agreed to allow derogations or exemptions from the directive on the grounds of serious economic and financial difficulties, experienced or anticipated, arising from take or pay commitments accepted in one or more purchase contracts.

Member states must notify a decision to grant a derogation to the Commission who have four weeks to endorse or reject the decision both using a list of criteria in the directive to reach their decision. Experience of the application of this Article will be reviewed five years after adoption of the directive.

UPSTREAM ACCESS

Member states must ensure that natural gas undertakings and eligible customers are able to obtain access to upstream pipeline networks, including technical services, except for the operational parts of such networks and facilities at the site of a field where gas is produced. The special economic, technical and operational characteristics of the upstream are being taken into account. The directive lists possible criteria for refusal of access to upstream pipeline network, such as difficulties, which could prejudice current and planned future production, including from marginal fields, and the needs of the owner or operator.

The treatment of access to the upstream became a major negotiating point for the UK, one of the few Member states with substantive indigenous gas reserves, and UK officials proposed the approach on which the agreement was reached. Certain provisions of the current voluntary Offshore Infrastructure Code of Practice would become statutory requirements in implementing the directive into UK law.

EMERGENT MARKETS AND REGIONS

Member states not directly connected to the interconnected system, or reliant on one supplier for over 75% of gas supply, can derogate from the directive for as long as these conditions apply. Only Greece and Portugal currently qualify as emergent markets. Member states may also apply to the Commission for temporary derogations where implementation of the directive would cause substantial problems in a geographically limited area of a Member state, in particular concerning the development of the transmission infrastructure. A derogation may only be granted by the Commission, against a list of criteria, if there is no gas infrastructure in this area or if there has been for less than 10 years. An example in the UK would be Northern Ireland. The temporary derogation may not exceed ten years after the first supply of gas in the area.

KEY ISSUES IN THE DIRECTIVE

It is still unclear how some countries plan to implement the Directive, and how far the objective of establishing a competitive European gas market will be met in practice. One of the key considerations concerns the rules governing third party access. The directives for both electricity and gas offer member states a choice between regulated and negotiated third party access. In electricity, most have opted for regulated third party access, driven by pressure in increasingly competitive electricity markets to establish power exchanges, for which regulated, or at least transparent and guaranteed network access are a prerequisite. In gas, most member states are contemplating negotiated third party access.

Beyond the legislative implementation of the directives, the operational detail will also be extremely important in determining the degree of competition to which the directives will lead. For example, whether third party access charges are distance-related or not will determine whether competition is more regional or Community-wide. While the headlines have been about third party access and access to networks, effective competition cannot come about without access to gas storage on competitive terms.

IMPLEMENTATION

The UK is the furthest advanced European state in terms of liberalizing its gas market with regulated third party access. Competition extends right down to the level of the domestic consumer and the UK has developed thriving spot, forward and futures markets. The UK privatization process began with the Gas Act (1986) and break up of British Gas' national market monopoly on supply.

Elsewhere in Europe, Spain opened 46% of its market in 1998 and plans to complete the process by 2013. The Netherlands aims to open 45% of its market by the end of 2000 and complete the process by 2007. Belgium opened 47% of its market in 1999 and proposes full market liberalization by 2010. Ireland already allows third parties access to 75% of the market and has further legislation in progress. At the other end of the scale France has been slow to implement the EU legislation and plans to open up its gas market by only the minimum levels (20%) required by the Directive in order to give state monopoly Gaz de France necessary time top develop its competitive position. This strategy mirrors its reluctance to wholeheartedly embrace the ethos of the electricity Directive. France expects to have a joint regulator for gas and power and plans to sell off all transmission pipelines to their operators Gaz de France and Elf, part of the TotalFina Group Elf.

EMERGING HUBS

The development of regional or European hubs could be seen both as a symptom and as a contributing agent in the spread of competition within Europe. In this sense hub is used to mean a trading and pricing centre rather than simply a collection of pipeline intersections. The UK's main hub is in fact a notional site rather than a physical and geographical location, the National Balancing Point (NBP) within the National Transmission System. This replaced Bacton as the delivery point for most physical transactions when claims validation problems at Bacton emerged in the mid 1990s.

Hubs can be used to generate prices that are used as benchmarks for other transactions. This facilitates both trading and risk management, and can help to create a truly international market. In continental Europe Distrigas has developed standardized terms and conditions for physical transactions based at Zeebrugge in Belgium and in addition has created a separate company Huberator N.V. to manage operations at the hub itself. At present Huberator is offering essentially a simple book keeping service tracking in and out flows of gas into the Hub, and not offering balancing or storage facilities. This means that gas arriving at the Zeebrugge Hub will have to meet the Distrigas specifications to be accepted. Any quality or continuity problems relating to gas travelling via the Interconnector that links the UK with continental Europe will have to be negotiated between the shipper involved and Distrigas. No standard Hub tariff will be used as some had initially hoped. In Austria OMV is marketing Baumgarten as a potential hub and other potential locations include Waidhaus, Rehden, Emden and Aachen in Germany and Dunkirk in France.

THE FUTURE

The benefit of experience in opening up the European electricity market will clearly assist the gas market. If the successful UK model (see box) is adopted within the remaining EU Member states then there is real potential for a robust convergent gas / electric market. The importance of this is born out by the progressive increase in gas-fired power generation at the expense of thermal generation. The progressive EU states in gas liberalization mostly mirror those electricity—UK, Netherlands, Spain—and it is important that the lessons learned in the developing electricity market are implemented in the gas market. And with competition in place, the trading platforms being developed for European electricity will be able to provide platforms for both gas and electricity. But while the framework for a competitive European gas market is now in place, with the opportunity for a convergent gas / electric market, the realization of this market is dependent on the individual EU Member states.

THE UK—A CASE STUDY

The UK is of particular interest because it is there that competition has evolved fastest. Many consider the liberalization process really began with the privatization of British Gas in 1986, and has continued apace ever since, with the opening up of the within day market in 1999. Liberalization has led to price volatility, and with this has come risk and the proliferation of risk management techniques.

New entrants, gas marketers and shippers have entered a supply chain previously dominated by a single company, British Gas. The introduction of competition in supply has meant that individual suppliers have to work hard to match precisely daily demand to daily supply, resulting in the development of a very short-term mechanism to provide for system imbalances. Gas typically demonstrates very high price volatility compared with oil or indeed other non-energy commodities, and this is even greater at the front end of the forward curve.

A real spot market began to emerge at the beginning of 1994. At this time most deals were concentrated around the one-month forward position (in 1994/95, for example, an estimated 85% of deals in the OTC market were for periods of one month forward, according to EJC Energy). Today activity is still concentrated at the prompt end of the forward curve but there is much greater variety of time period available. There has been a shift towards hedging specific periods, for example particular weeks or weekends.

Currently there are approximately 60 buyers and sellers active in the UK gas market. These include producers, shippers and gas users such as industrial consumers and power generators. In addition there are about 12 over-the counter (OTC) brokers. The risk management techniques that exist in the OTC market are forwards, swaps and options. These deals are monitored and reported by a number of different price reporting services which also cover spot markets. It should be noted that the number of deals done does not necessarily equate to the amount of risk management activity. Speculative use

THE UK—A CASE STUDY (continued)

of the new tools has also risen, with the high price volatility creating potentially attractive opportunities.

In January 1997 the IPE launched a monthly natural gas futures contract physically deliverable at the National Balancing Point (NBP). It had been originally designed to be based on Bacton which was the centre of physical trading. But once physical deals and therefore associated risk migrated to the NBP, the IPE shifted the delivery point of the contract in order to meet the needs of the industry.

Since launch the contract has been adapted to suit new industry requirements. Its scope has been extended to three years and daily contracts have been added. Plans for the futures include quarterly contracts, spread and strip trading facilities and the introduction of a within day contract. By the end of its third year of operation the IPE contract was trading about 2000 lots a day, a volume equivalent to just under total daily UK natural gas consumption.

The contract is traded on the IPE's Energy Trading System (ETS), which has also experienced extensive upgrading over the last three years. An international communications link has been established with Nord Pool, the Nordic power exchange which runs electricity spot forward and futures markets in northern Europe. This link will ultimately be used to enable Nord Pool users to access IPE markets, and IPE users to access Nord Pools markets. With gas and electricity showing significant synergies it is hoped that this link up will provide a useful service to the wider European marketplace as well as assisting the IPE to benefit from Nord Pool's experience when developing new contracts for the UK electricity sector.

The IPE's natural gas market offers the usual advantages of futures contracts. Not only is it a liquid market, where it is very difficult for one trader or a group of traders to influence price, but also it is fully transparent. Prices for individual deals are transmitted in real time to approximately 17000 screens in 45 countries (check). This has led to the contract being used as a benchmark in physical or

THE UK—A CASE STUDY (continued)

other over the counter transactions. Short and long term deals have been transacted using either the settlement price or the Natural Gas index

The October 1998 completion of the Interconnector gas pipeline that links Zeebrugge in the UK with Zeebrugge in Bacton offers the possibility that a UK hub, either NBP or Bacton could be used for continental European. At present it is too early to say what the outcome will be but the development of hubs is an important step towards the development of a truly competitive and integrated European gas market.

CHAPTER 8

Coal Trading: The Core of Change

By Louise Croucher and Alan Gillespie

A liberalized and deregulated electricity market place has already started to create enormous price competition among generators and it is inconceivable that this situation will not affect the steam coal market. It is expected that during the period 1995 to 2020, demand for energy worldwide is forecast to increase by 75% and fossil fuels are expected to meet 95% of this additional energy demand. The coal market's sentiment to change has been one of reluctance, however the economics of power generation suggests that generators will be financially drawn to the cheapest practical fuel source and will acquire and manage that fuel in the manner most likely

FIGURE 8.1

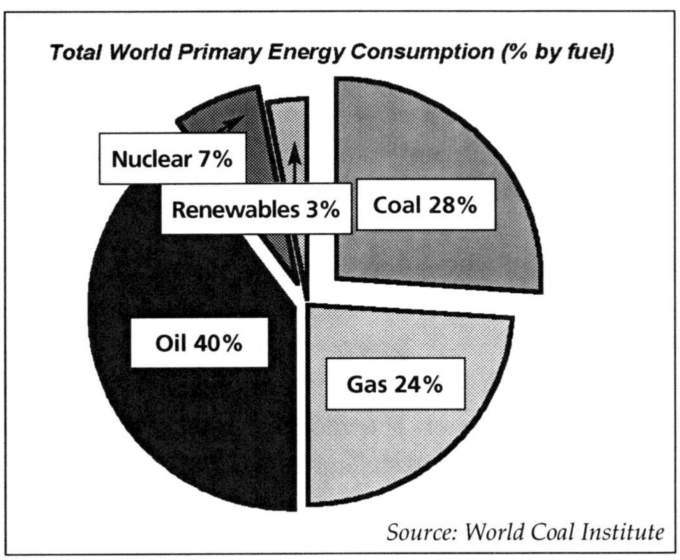

to protect their generation margin. This will, and has already started to, change the opinions of coal buyers and sellers worldwide.

The worldwide trade of coal can be broadly split into five geographical areas: North America, South America, Europe, Africa and a combined Australia and Asia market commonly referred to as Australasia.

It is both interesting and significant to the international coal market that the largest coal producers in the world are not necessarily the largest exporters. The result of this 'tip on scales' makes for a market unique in both its size and diversity. The Peoples Republic of China is the largest producer with 1236* million metric tonnes of coal being produced in 1998.The US is second with 936* million metric tonnes and in both cases the majority of production is consumed within its domestic borders. For some countries, such as Australia and Colombia, coal exports are of much greater importance to the national economy than the domestic consumption of their coal. The three largest exporters are Australia, South Africa and Indonesia.

WORLDWIDE COAL MEETS CHANGING ENERGY MARKETS

The deregulation of global electricity markets has revolutionized electricity from being a utility service product into a 'commodity'. Utilities, now looking to compete on a larger scale, are taking advantage of alternative ways to buy physical fuel and participate in risk management opportunities that are available throughout the world. The OTC electricity market has evolved in the UK since 1991 and in the US since 1994. Deregulation in Continental Europe is ongoing, as member states have opened up 25% of their domestic market to competition since February 1999. Australia is experiencing steady progress and has had an OTC market in place since 1998.

Prior to the electricity deregulation that is currently accelerating around the globe, utilities traditionally bought coal under long term contracts, which secured supply to their 'guaranteed market'. Of course there will continue to be a market for long-term coal contracts and locking in coal prices into the future. However in recent times there has been a substantial move towards shorter dated contracts and spot business.

The correlation between coal and electricity is fundamental in the understanding of the changing attitudes towards coal risk management.

FIGURE 8.2

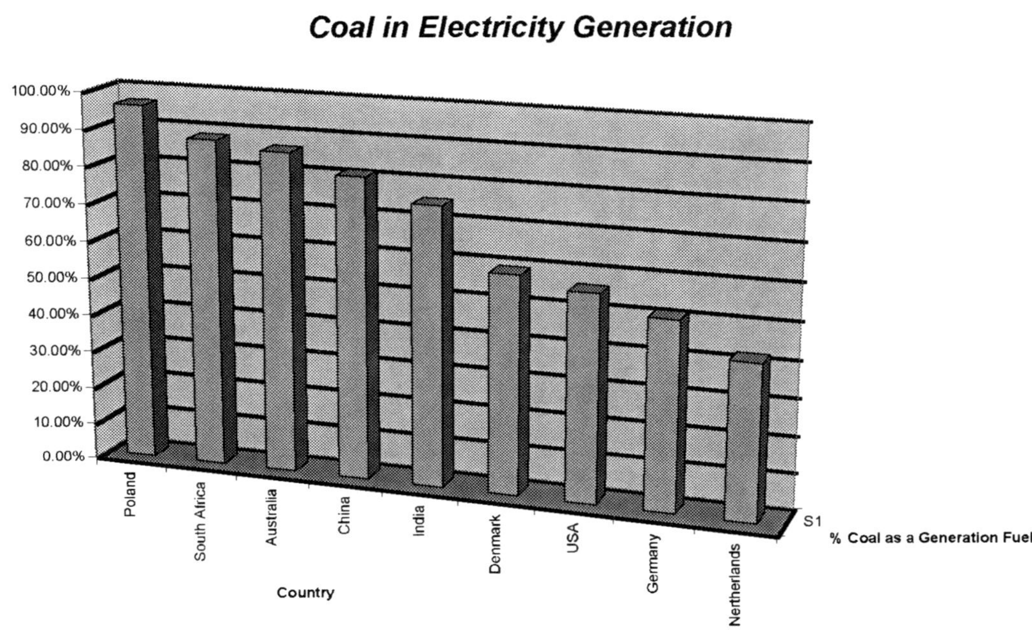

Source: World Coal Institute

NORTH AMERICA

In the US, every $1 billion worth of coal production generates approximately $19 billion worth of electricity. Since Title IV of the Clean Air Act was passed in 1990, domestic utility coal usage has increased by 18% and as a result more than 56% of US electricity is now generated from coal-fired power stations, utilizing 87% of US coal production. [lrc1][lrc2]In 1998, steam coal exports from the US only amounted to 3% of its total production with this percentage expected to decrease. The majority of these exports found their way into Canada, with the balance being split between Asia and Europe.

The length of coal contracts in the US have been substantially reduced from 20 to 25 year deals that used to be commonplace to the current, and more popular, one to five-year contracts which include

FIGURE 8.3

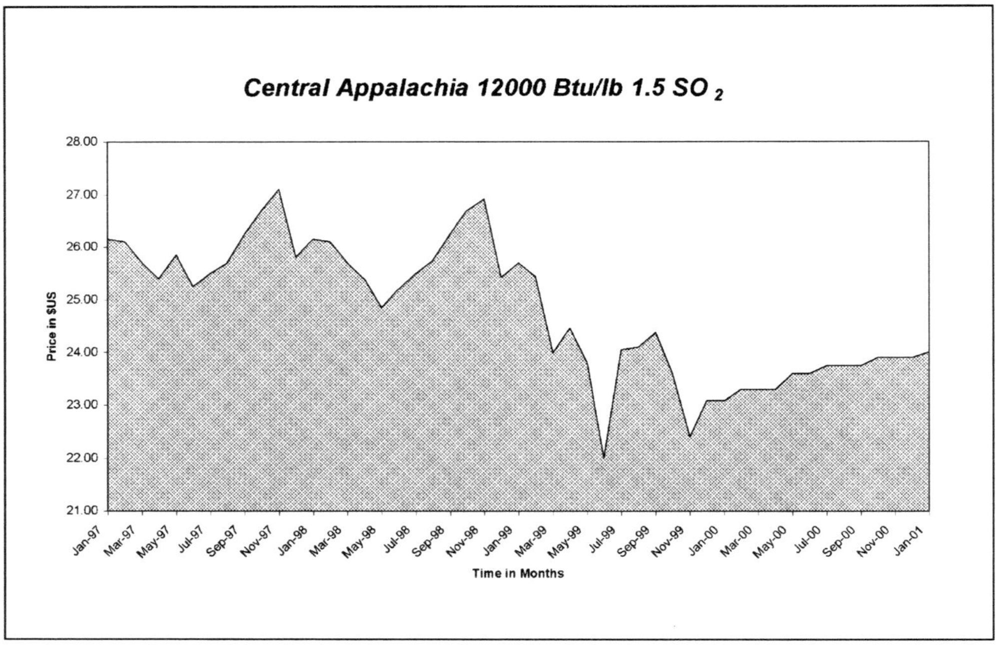

annual re-negotiation of price. Again this can be attributed to the increase of competition among generators and the changing mentality of the coal industry. Due to the collapse of world steam coal prices, export contracts from the US to Europe are not being renewed as they expire.

The US steam coal market is divided into three primary regions:

- Central Appalachian coal is the predominant coal within the **Eastern Region**, providing 240 million metric tonnes or 62% of the regions supply base. This coal has a Btu/lb. value ranging from 11,500 - 12,500 and sulfur from 1.2 - 2.0lbs. Within Central Appalachia, the Big Sandy River provides origination for the many power plants within the Eastern market. The coal is transported via barge to large utilities on either the Big Sandy or Kanawha River, and by rail on either the CSX or Norfolk Southern railroad.

- The **Interior Region** is in decline, supplying just 15% of coal to the US market.

- Powder River Basin (PRB) makes up the lion's share of the **Western Regions** coal production, accounting for 324 million metric tonnes, or 73%, of the regions supply base. There are two main grades of PRB coal: 8,400 Btu/lb, 0.8lbs sulfur dioxide; and 8,800 Btu/lb 0.8lbs, sulfur dioxide. PRB Coal is mined in the Wyoming region of the US and is transported on unit trains on the Burlington Northern Sante Fe (BNSF) Railway and / or Union Pacific Railroad. Both railroads have the capability to transport coal into markets located across the Northern Central, Midwestern, Western and Southern regions of the US.

US 'POWER HUNGRY'

The US is the largest producer and consumer of electricity in the world. Its annual net energy for load is approximately 3.5 billion megawatt hours, representing a major component of the US economy. Electricity has generally been supplied by large, heavily regulated, companies or publicly owned utilities with defined territories, fully regulated pricing and authorized rates of return.

The National Energy Policy Act of 1992 (EPA) was a needy catalyst to speed up the deregulation process of electricity in the US. The EPA greatly increased industry competition and gave the Federal Energy Regulatory Commission (FERC) authority over transmission assets and owner operators.

The most important element of the EPA was its efforts to separate, or 'un-bundle', generation, transmission and distribution. It effectively established open access to the power transmission grid by all qualified participants being only subject to physical constraints. In 1996, FERC introduced orders 888 and 889 resulting in the development of power pools around the country and the active encouragement of independent system operators to create a clear separation between generation and transmission businesses.

Unfortunately, while wholesale deregulation has been hastened by the aggressive rulings of FERC they have little or no control over the next step of implementing retail open access as, with a few exceptions, this is governed by individual states. Political and other action groups have hindered efforts to change, preventing consumers from choosing their provider. Some states have even passed legislation that has put in place barriers to trade competitively. The disappointment within the industry is apparent. Most of the major investments in developing assets and trading entities were primarily in anticipation of a liberalized retail market. Most concede that either by progressive state legislation, or more heavy-handed federal guidance, there will be universal retail trade of electricity before 2010.

SOUTH AMERICA

In South America, Colombia and Venezuela are the primary sources of coal and will continue to play an important role in the world coal market because of low production costs and large volumes of un-mined coal in the region. In Columbia, producers have the potential to significantly increase future coal exports with this increase partly dependent on the new owners of Carbocol (a main state owned producer) after privatization.

A number of foreign investments have also played an important part in the growth of South America's coal interests. By providing the necessary infrastructure, the problem of transportation required for exportation is being addressed and resolved.

But despite its advantages, the demand for Colombian coals has decreased in parts of the world, as consumers have been able to obtain supply from more competitive sources. Poland, Russia and, to a lesser extent, South Africa are satisfying consumption needs more steadily in Europe.

EUROPE

Traditionally, the European coal market traded around long term supply contracts with producers and consumers locking in prices from one to 10 years forward. Transactions were typically negotiated annually with no risk management strategies in place up until recently, where a slow transition towards more spot business has taken place and more hedging possibilities are being explored. The volatility of prices within the European coal market has increased and recent trends expect this to continue in the future.

The position of Amsterdam, Rotterdam and Antwerp (ARA) and the capacity of the ports have made ARA a major shipping area for coal into Europe. The trading of ARA stockpile coal has seen significant changes over time. Historically, a major trading hub for coal ARA recently experienced a decline in stockpile tonnages but these are expected to increase in the future.

Coal is shipped into the ARA regions predominately from the large coal exporting nations of South Africa, Colombia, Australia and Indonesia. The coal is transported in panamax, or capsized, vessels and then broken down and delivered in smaller barges and coasters within Europe.

Traded coal from an ARA stockpile used to be an integral part of coal trading in Northern Europe. Shipments of coal were chartered with either no firm customers, or with only part of the cargo sold, discharged on the quayside and customers then found. However, this aspect of the business environment has declined almost to the point of non-existence. Currently depressed coal prices have made it non-productive to store coal at a terminal stockyard, with best-cost estimates of $4 per tonne to discharge and then reload on to a coaster. On top of this, if the pile were left for any duration, extra storage costs would have to be factored in to the equation. Take this against a normal discharge cost of under $2 per tonne and you are looking at least at an extra cost of 14% on the present ARA CIF (Carriage Insurance Freight) price.

Shipping reports show there are coasters full of coal moving from the ARA ports to the UK and around the European coast. The geographical trend has changed recently with increasingly competitive prices and coal being exported from Poland and Russia. Trading from ARA does exist, and is expected to increase in the near to medium term. The continued supply of small coasters from the former Soviet Union and Poland into the European market should see the re-emergence of an active ARA Barge market, as large global players are forced to remain competitive to ensure continued sales of the underlying commodity.

EUROPEAN ELECTRICITY

In Europe the continuing deregulation of the electricity market is the main driving force behind a commoditized ARA market. The privatization of the industry in England and Wales through the Electricity Act (1989) gave birth to an open commodity market: the Pool. The UK, Germany, Norway and Sweden have already opened up their markets to external players. Other European countries are following suit although the market structures vary considerably from one country to another. The implementation of the European Union's Electricity Directive in February 1999 promised much change for consumers of electricity in Continental Europe as member states opened up 25% of their domestic market to competition. Some member states, such as Germany took greater steps and opened up 100% of their market to competition, resulting in substantial trading volume of electricity for physical delivery.

There was initial resistance to the directive as utilities strived to protect their established market position and viewed the changes as a threat to their profit margin. Utilities were exploring merger opportunities within countries to create a stronger force to contend with cross border competition. The changes are seeing generators having to be financially drawn to the cheapest practical fuel source and looking at a number of ways to protect their generation margin.

AUSTRALASIA

The Peoples Republic of China is the world's largest coal producer and recently has gone through major consolidation on the supply side. China's export trade is marginal, the majority being used for their own consumption however with their low costs of production they could become a major force in the export market, supplying low cost coal into this area.

Indonesian producers should see an increase in their exports, as consumers become more accommodating towards the burning of sub-bituminous coals. In India, because coal is of low quality it does not lend itself to international trade and is used predominately for domestic purposes.

Australia is the largest exporter of steam coal in the world with 80* million tonnes of steam coal being exported in 1998. New South Wales is the dominant coal-producing region within Australia, exporting the majority of its coal through the port of Newcastle. The Far East is dominant market for Australian coal, although in the latter half of 1998, and the first quarter of 1999, the Australian presence in the European market increased tremendously due to the collapse in the freight market. The recent tightening of freight rates however has reversed this flow, and once again little Australian coal finds it way into Europe.

AUSTRALASIAN ELECTRICITY: GROWTH POTENTIAL

Changes in Australian electricity began in 1991 with the development of the National Electricity Market (NEM). The Industry Commission recommended to the federal and state governments that introducing competition in generation could increase power output and retail supply to customers. Individual state governments and territories have introduced their own reforms in preparation for the NEM and to restructure the industry into a competitive privatized market. The state of Victoria has privatized and unbundled its vertically integrated electric utilities and other states are in the process of following its lead. Although The Sydney Futures Exchange (SFE) lists electricity futures it has little impact on overall market trading.

The New Zealand market is fully deregulated but due to its size has little impact on the worldwide stage. The New Zealand Futures and Options Exchange, which has an alliance with the SFE, also lists electricity futures. In Japan a few IPP (Independent Power Stations) have been built and a limited number of tenders have been put to the market for supply but there has been no government initiative towards market deregulation.

AFRICA

South Africa dominates coal trade on the African continent. Exporting approximately 58* million metric tonnes of steam coal each year it is the second largest exporter of steam coal in the world. The Richards Bay Coal Terminal (RBCT) is the main coal facility in South Africa accounting for over 85% of their exports. Europe is South Africa's main customer with approximately 60% of all exports being shipped into this region. This trend is set to decrease as more competitive coal finds its way into Europe from Poland, Russia and potentially Colombia. As a result the South African producers have turned their focus on supplying increased tonnages into the East with particular success being achieved within the Indian market.

South Africa is very reliant on coal for its own electricity generation. Coal fired power stations provide almost all of the countries power needs.

COAL DERIVATIVES: DEVELOPMENTS AND IMPACT

With electricity deregulation in full swing the timing was ripe to introduce coal derivatives as a means to hedge this vital generation feedstock fuel. Initially, studies focused primarily on the development of a coal futures contract, which proved premature due to the lack of liquidity and volatility in the marketplace. A more suitable solution revealed itself in the form of coal swaps, a derivative that moves at the same pace as the underlying commodity. The introduction of coal derivatives into the market is attributed to over the counter (OTC) brokerage Tradition Financial Services (TFS), which set up a coal desk in August 1997.

The group was instrumental in the educational development of this market with seminars held in Europe, the US and South Africa. At the beginning it became apparent that the coal market needed educating of this new concept. The definition of derivatives and the difference between futures and the OTC market had to be conveyed before the benefits that these products provide could be explained. Terms and scenario's that were taken for granted in other energy related commodities were clearly a foreign language to the coal industry. The seminars focussed on the basics of derivatives and the flexibility of the OTC market, the changing energy markets and its impact on the coal industry and different deal structures, particularly swaps, and the theory behind using them.

Early in 1998 TFS conceived the first ARA basket index—the TFS API#1—as a reliable reference point for swaps. Prior to this the European market had no established benchmark index. The TFS API#1 is a monthly basket index for the ARA CIF coal price (basis 6000 Kcal/kg NAR). The prices used for this index are the FOB barge quotes published by *Coal Week International* and *South African Coal Report* along with the *MCIS European Steam Coal Marker* published in the *International Coal Report*. The three constituent parts of the index are actually sourced from different sectors of the market and yet history shows that they broadly reflect the same price trend.

In April 1998, TFS and the OTC coal derivatives market proved the skeptics of coal derivatives wrong by trading the worlds first financially settled coal swap. The deal was based on 6,000 Kcal/kg ARA CIF NAR coal and used the TFS API#1 as the benchmark index. Coal producers, traders and consumers were now able to hedge against adverse price movements

FIGURE 8.4

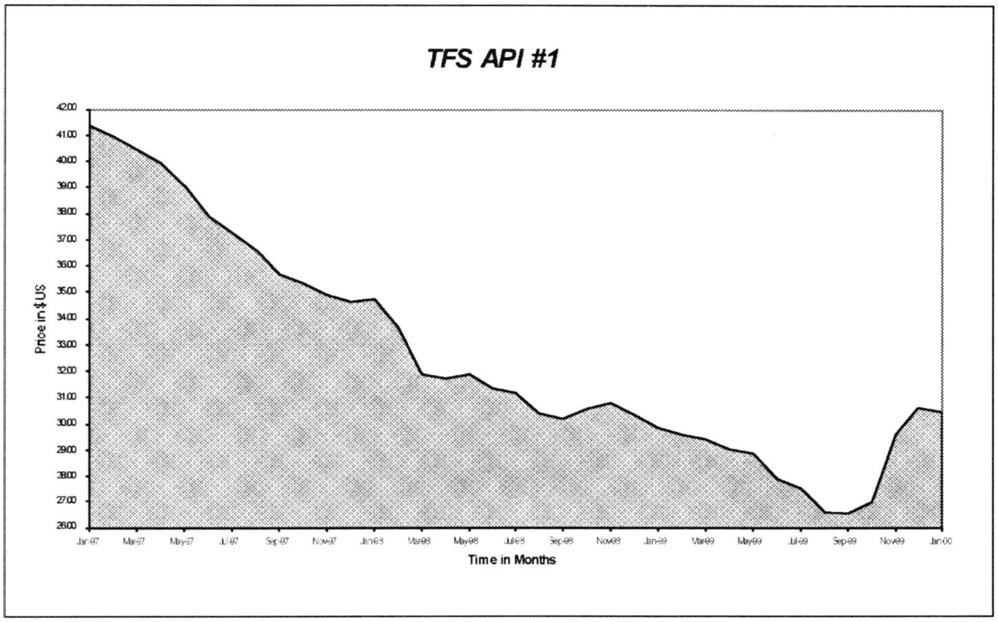

and so 1998 began a year of firsts in the coal derivatives market. This included the first US coal swap in June, the first modern floating physical transaction in November, and the first petcoke swap in December.

On the heels of European success, expansion of the OTC market into the US was met with a favorable response. Traders and brokers alike saw the addition of these structures as an enhancement to the long-standing physical market and began trading both physical and paper transactions in earnest.

The Eastern region saw significant increases in market volume primarily due to the development of a futures contract by the New York Mercantile Exchange (NYMEX). While the contract has been established but not yet launched the industry is already utilizing the NYMEX terms and conditions. The Western US market is fairly liquid but still emerging and currently trails the Eastern US market in overall volume by approximately half.

The US emissions market also benefited from coal derivatives as TFS brokered the worlds first coal / emission option in May 1999. For the first time ever participants of a coal transaction were able to settle any differences between the delivered sulfur content and the agreed benchmark in emission allowances. This deal was noteworthy in two further respects: first, it quantified the sulfur content of coal at an absolute market value; and second, it extended the bundling potential of coal and emission allowances beyond the simple `barter` deals the market had so far seen.

To allow for continued choice and flexibility August 1999 saw the birth the TFS API#2 basket ARA index. The prices used are the *MCIS European Steam Coal Marker* published in the *International Coal Report* and the *International Index* published monthly in *COAL Daily*.

FIGURE 8.5

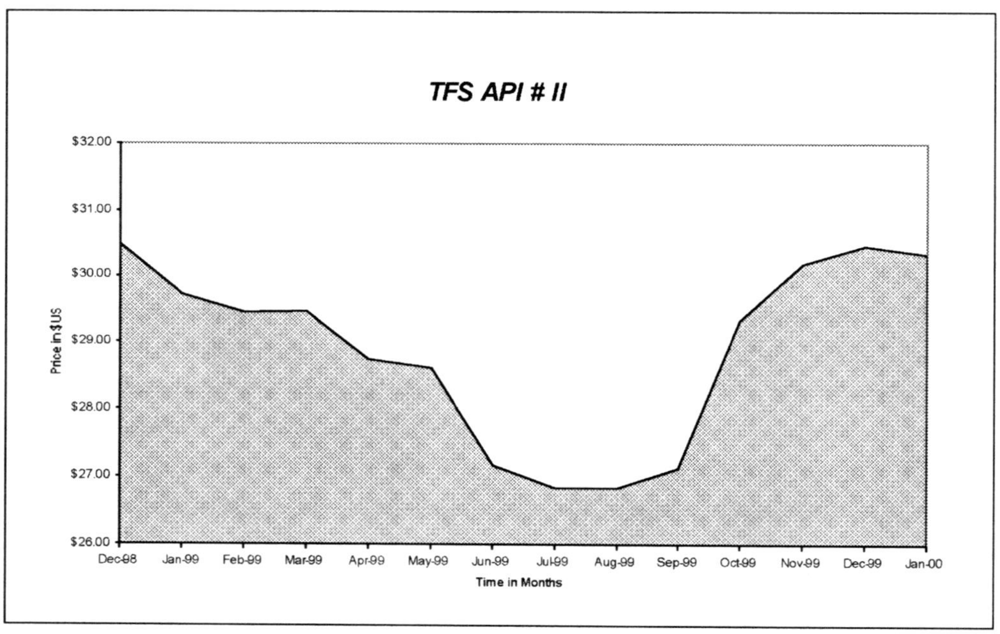

The newest development in the European steam coal market is the development of the Standard European Coal Agreement (SECA). It is a standardization of previously proposed contracts combining in-depth market research with professional legal consultancy. The terms and conditions of the contract have been formed by a number of coal industry participants and legal representatives. SECA has been launched primarily as a physical ARA FOB coal contract but, like the NYMEX look-alike in the US, participants will have the ability, where mutually agreed, to close out the position should neither party want to make or take physical delivery. TFS was pleased to conclude the first physical option under SECA in January 2000.

TABLE 8.6

SECA SPECIFICATIONS

The Shipment shall comply with the following Specifications determined in accordance with ISO standards on an "as received" basis:

CHARACTERISTICS:	TYPICAL:	REJECTION LIMITS:
NCV:	6000 Kcal/kg	less than 5800 Kcal/kg
Total moisture:	13%	more than 15%
Ash:	16% (max)	–
Sulphur:	1.0% (max)	more than 1.0%
Volatile matter:	24-38%	less than 24% or more than 38%
Hardgrove grindability index:	45 - 56	–
Size range:	0 x 50 mm	–

Practically speaking, a working coal derivative market provides a tool to enable coal market participants to switch between fixed and floating depending on their view of the market. Indeed some way down the road the use of coal and electricity derivatives will enable the kind of total flexibility that traders in the oil markets take for granted.

As the market witnessed the beginnings of the implementation of professional risk management techniques the importance of coal benchmarking became more apparent. Coal companies were realizing that meaningful indices were needed to `mark-to-market` physical and derivative coal exposures on a more frequent and accurate basis.

As sophistication in the coal market grows, the focus on published indices will continue to sharpen. The coal indices published by the top coal trade publications could eventually become as important to the coal industry as Platt's prices are to the oil business. Many will know Platt's as the benchmark for the majority of oil transactions, swaps and physical.

There have been a few reservations from certain coal trade participants regarding the validity and robustness of published coal indices and the market has, and is continuing to, apply pressure to these publications to improve their pricing techniques. Since the development of coal derivatives editors have realized just how important their price tables actually are.

On the back of power deregulation and other modernizing forces, the industry is already finding it useful to link physical deals to an index in the way that they are in so many other commodities. The 'floating' deal, 'floating' or 'indexation' is a simple technique where coal deals are linked to an index and float with that price.

As we enter the new millennium, energy companies need to be at the forefront of competitiveness to add value within the new marketplace. The role of the developing OTC market is to provide market players with the necessary tools to achieve these unlimited opportunities that now present themselves. There is no turning back from here. Coal's main customer, power generation, is changing. Coal is not far behind

COAL DERIVATIVES AND RISK MANAGEMENT STRATEGIES

There are so many opportunities in the market right now and so many deals to be done. Coal swaps, coal linked to power, gas, fuel oil, weather, emission allowances, coal that floats with the spot price, coal supply that is interchangeable with other fuel sources, generators re-selling, producers buying and re-selling, using coal as cash in re-financing, more creative blending, time arbitrage, sulphur and ash arbitrage, tolling.... The list goes on. Below are a few basic strategies applied on an ongoing basis today:

- **Swaps.** The swaps that are being traded in the coal industry are plain vanilla fixed for floating transactions, settled against an industry standard index on a predetermined settlement date, depending on the duration of the deal. This basic model has brought flexibility to fixed price contracts and enables participants to hedge and speculate effectively without being obliged to make or take delivery of physical coal. Utilizing the OTC market allows financially settled swaps to be very flexible instruments; the price, quantity traded, the period over which the deal is made and the basis on which it is priced are all open to negotiation and agreement between the counterparties. A collar clause can be written into any coal swap deal to limit the profit and loss profile of the trade to a predetermined figure. These tools can be used to hedge the exact quantity of a physical contract or be used as a pure speculative trade.

- **Floating Physical Transactions (Indexation).** This is a transaction where physical coal is bought and / or sold at a price that floats against a chosen index.

- **Spark Spreads.** A trading strategy involving a coal swap and an electricity swap, designed to reduce the risks associated with a generators operating cash margin (input / output).

- **Price Linkage.** An extension of indexation (floating physical), coal trades can be linked to power indices. For example the monthly average electricity price determines the amount the generator pays for its coal.

- **Time Spreads.** A trading method whereby participants can take advantage of the differential in the coal prices between two different time periods, most commonly being a quarterly spread.

- **Options.** Although options have been used within the coal industry for many years, these options have traditionally been offered at no cost to the holder thereof and are generally limited to a single consumer. Producers, therefore, receive no compensation for taking on the additional risk of being uncertain of their actual supply / demand position.

- **Nymex Look-alike.** The OTC adoption and utilization of NYMEX terms and conditions. The trades are in barge lots of five barges, minimum 7,750 tonne increments with an option to take physical delivery or book out the position.

- **Bundled Transactions.** A coal / emission transaction allowing participants to settle any differences between the delivered sulfur content and the agreed benchmark in emission allowances.

- **SECA (Standard European Coal Agreement).** A physical ARA FOB coal contract with the ability to close out the position. The contract is traded in lots of 5,000 metric tonnes for delivery, seller's option, FOB ARA. The pricing will be on 6,000 Kcal/kg NAR FOB ARA adjusted pro rata.

CHAPTER 9

Electronic Energy Trading

By Peter C. Fusaro and Jeremy Wilcox

Electronic Commerce (E-Commerce) opportunities for energy trading are continuing to evolve and grow. This chapter attempts to assess developments of electronic trading platforms for energy trading, both for exchange-traded futures and over-the-counter derivative instruments in rapidly changing markets. But it is only a snapshot of a dynamic situation. The energy industry is significantly conducive to the use of risk management tools due to its underlying price volatility, and therefore a logical candidate for the use of an array of electronic trading platforms. Price discovery risk will be enhanced using the technology of the Internet.

Electronic commerce is transforming energy markets. The mature markets of oil and gas trading as well as the emerging markets for electric power, emissions and weather trading are ripe for trading on electronic platforms. This chapter provides regional analyses of market developments (North America, Europe and Asia Pacific) and implications for the ancillary markets of emissions, weather and bandwidth. It explores the market drivers for the changing face of energy trading from open outcry to screen trading for price discovery.

THE EVOLUTION OF ENERGY TRADING

Energy trading began after the end of Official Selling Price (OSP) programs by the major oil companies and OPEC nations after the 1973 Oil Embargo and coincided with the development of a spot market for crude oil and petroleum products. In 1978, the changing structure nature of the physical spot market for oil presaged the development of energy futures with the successful launch of the New York Mercantile Exchange (NYMEX) heating oil futures contract, which was tied to its physical delivery in New York

Harbor. Successive oil futures contracts and the development of an active Over-the-Counter (OTC) market for forward oil trading in the early 1980s brought significant structural changes to the international oil industry. In effect, price transparency accelerated both physical and financial trading of crude oil and petroleum products globally. In April 1990, the NYMEX launched the very successful Henry Hub natural gas futures contract, which simultaneously coincided with the development of an active OTC natural gas market.

Electricity trading began with the Nord Pool contracts for the Scandinavian markets in 1993. NYMEX, the Chicago Board of Trade (CBOT) and the Minneapolis Grain Exchange (MGE) have since launched eight failing electricity futures contracts. In this case, the OTC market for electricity derivatives in the US began in late 1993 prior to the futures contract launches which began on March 29, 1996. Clearly, something had changed. The underlying retail and wholesale US electricity markets are valued at over $300 billion and growing. It indicates a potential financial electricity derivatives market of over $2 trillion. But electricity futures have not and will not work in the US, and there are doubts that it will work in continental Europe. What had changed is the structure of energy futures trading. The age of electronic trading coupled with OTC market flexibility have usurped exchange-traded electricity contracts. The exchanges have been slow to react to this phenomenon.

Other critical changes have occurred over the past twenty years, price assessment panels and index trading, which failed in the late 1980s, are succeeding in the 1990s. A sea change in energy trading is underway. Electronic index construction coupled with screen trading is already changing the industry globally. Electronic broking and trading platforms are emerging that will continue to change the face of energy trading.

Ironically, while these factors come into play, oil futures and natural gas futures trading have grown substantially during 1999 with NYMEX WTI crude oil futures (its benchmark contract) growing over 50% in open interest this year and average daily trading of 149,000 contracts per day. Subsequent growth has been exhibited on the International Petroleum Exchange (IPE) Brent crude oil contact trading 64,000 contracts per day in 1999. But new commodity markets for energy are also emerging for weather and emissions trading. These will not necessarily be financial futures since they are trading on OTC markets and may be conducive to the wave of e-commerce that is enveloping the global energy trading community.

Changes underway in energy trading are impacting on this capital intensive and conservative industry. The energy industry is on the brink of dynamic and dramatic fundamental change both in the physical and financial markets around the world. Electronic energy trading is now emerging across the globe in all energy markets. Companies such as Altra Energy Technologies, Houston Street, Swapnet, Bloomberg, RedMeteor.com, PEPEX (Petroleum Electronic Pricing Exchange) and the like are leading the way. The futures exchanges are beginning to face this global challenge. For the energy business is consolidating, restructuring and concentrating on a larger scale like never before. Margins are razor thin therefore volume becomes the only game in town and the need to move more barrels, molecules, or electrons is paramount.

It should always be remembered that energy hedging is still in its infancy with most producers and consumers not hedging. Deregulation shifts more risks to companies so that more trading and hedging is inevitable. Moreover, the technological drivers of electronic trading and the Internet will fundamentally change the structure of energy markets that will inevitability enhance market liquidity across the energy complex and around the world.

Business-to-business e-commerce is already becoming a major part of the global energy trading markets and has been estimated by Forrester Research to grow to $266 billion by 2004 including online exchanges, auctions and retail aggregators. It has been estimated that electronic trading of natural gas, which has been active over the past four years, accounts for about one-fourth of the entire US Internet economy. Electronic trading of gas, power, crude oil and natural gas liquids is already underway and allows a selection of creditworthy counterparties, which is a major concern in energy markets since many energy companies have lesser credits. Electronic trading also can reduce transaction costs through greater economies of scale, an advantage over both futures exchanges and brokers. Electronic energy trading will also be integrated into a robust price risk and transaction management system so that real time trading operations can be integrated into a company's front to back office.

The next wave of electronic energy will be in the retail markets as customer choice initiatives take hold through further deregulation. The ability to choose energy suppliers including energy measurement and bill payment through the Internet is just starting to take hold and is being offered by some utilities. The future will also integrate not only energy bills but telecommunication and water bills into one Internet-based bill. Secure payment will be made by credit card over the Internet. Other Internet applications will be brought forward in the form of aggregators, which is key to unlocking the power of retail markets because of diffusion of buyers and sellers. Fragmented markets create inefficiencies. Robust electronic exchanges are the next step in the transformation of the energy industry toward an e-commerce base.

Energy deregulation created the need for newer information systems that could support competitive markets. The Internet has become the tool required for the next generation of energy trading, which is faster, higher volume, and needs IT to be successful. This fundamental change will affect energy trading because of the dynamism of the global markets.

Paper market trading of energy requires viable cash markets i.e. the underlying physical market. This is not true of many paper markets. Today, there are viable cash markets for crude oil, residual fuel oil, domestic heating oil, propane, natural gas, coal and electricity. Some of these are global markets (for oil) but most are regional markets (for power and natural gas). Nevertheless, these physical markets are continuing to become commoditized. As these markets mature, trading margins become thinner and greater trading volumes are needed. Electronic trading will facilitate the commoditization of new markets, change the role of traders and brokers, and create more arbitrage opportunities globally for energy traders. It will create more price discovery, which is key for structured financial products, which need deal flow and data. It is also a driver for OTC market liquidity rather than futures trading.

One of the problems in the past for electronic trading systems was that they had been dependent on costly, dedicated private networks and computer hardware, which added overheads for users making them less competitive against conventional telephone trading. Now though it is possible to harness the power of the Internet for business applications and offer global business-to-business E-Commerce solutions for traders with no up front cost. Users only need access to the Internet.

The variety and scale of the electronic trading platforms would seem to indicate that there will be a large ramping up of many competitive systems, a consolidation period, and then the emergence of clear winners. Since many new and unknown competitors are in the offing, it is helpful to look at the existing systems of today and evaluate their road to success or failure. This discussion will include electronic exchanges, OTC brokers, and the development of e-trade capability by traditional floor exchanges such as NYMEX, IPE and the Singapore Exchange.

PIT TRADING

Pit-traded futures contracts are under siege globally by the rapid rise of screen trading, but energy markets in the US may be the last bastion of open outcry trading. The existing futures exchanges are trying to move to electronic trading without disrupting or cannibalizing the existing face-to-face market of open outcry. Thus, they have been attempting to offer electronic trading during after hours trading. This has been a stop-gap measure. The reality is that moving to an electronic trading format is becoming a very difficult process for futures exchanges to face up as oil and gas trading on NYMEX during 1999 has been robust and similarly oil trading on the IPE has had a banner year. It is hard to change a very successful and lucrative model.

However, the key financial market change was the shift of the Deutschmark from the London's LIFFE (London International Financial Futures Exchange) to Eurex in a manner of months when Eurex went electronic in early in 1999. LIFFE eventually went electronic but lost its momentum. This incident was a wake up call for the futures industry that electronic trading was real and an accelerating threat to the traditional monopoly of floor-based futures trading.

The second electronic competitive threat is proliferation of cheap electronic communication networks (ECNs) that are already threatening both financial and commodity exchanges. Unfortunately, exchange members are slow to adapt since they have an interest in maintaining the status quo and have been reluctant to move aggressively from floor-based to screen-based trading. ECNs match buyers and sellers without a need for voice confirmation. ECNs such as Island and Archipelago have already stolen volume from the New York Stock Exchange.

The question, thus, becomes how will electronic trading transform energy markets not when. Energy brokers are trying to forestall this event by pooling their gas and electricity data through PwC's indexes—Gastrax and Powertrax. But this is really a holding action of a 'broker-assisted' network that will fall by the wayside in the wake of rapid technological change. System openness will cause these alliances and closed systems to dissipate.

Their clients are not technologically phobic and will gravitate to new trading solutions based on ease of access, cost and reliability of the emerging system platforms.

TECHNOLOGICAL CHANGE AS MARKET DRIVER

Energy markets are conservative in nature and thrive on security of supply. The avoidance of risk would seem to be a curious place to foster the electronic future, but the added impetus of energy deregulation, as a global phenomenon, is bringing the technology solution to the industry quite rapidly as a consequence of more market risk. Liberalization is the process of introducing competition and brings with it radical changes to the structure of the industry. Traditional business practices tend to disappear, as new competitive forces are unleashed. Moreover, new competitors such as Oracle, Microsoft, AT&T, British Telecommunications and IBM already have made inroads into this industry for many years.

The key changes of liberalizing markets are changes in market share from monopolistic ownership, exposure to risks necessitating a new risk management infrastructure be developed for companies, and market consolidation over time. Liberalization also forces a new and very different commercial environment that brings forth new IT solutions since the business as usual model in the use of computer systems is now inadequate. The infrastructure transition from monopoly to competition focuses understanding on the need for new trading expertise, systems, and exchanges. It is definitely a new world affecting, market share, procurement patterns, and price volatility. The radical restructuring of the energy industries in oil, gas and power across the world is accelerating. New solutions are rife.

NORTH AMERICAN TRADING PLATFORMS

In the North American markets, there are several existing and new electronic energy trading platforms for oil, gas, power, and emissions. During the next two years, one or two systems should become dominant as the market enters a consolidation phase; however, today we have many choices and more to come. NYMEX ACCESS, Houston Street Exchange, Altra

Energy Technologies, Enron Online, Enermetrix, Cantor Fitzergerald, Bloomberg, RedMeteor.com, PEPEX, IntercontinentalExchange and Natsource are briefly examined in this chapter (greater detail is offered in our special report, **Electronic Energy Trading** www.global-change.com). It is not an inclusive list of all North American platforms.

- **New York Mercantile Exchange (NYMEX)**—NYMEX ACCESS (American Computerized Commodity Exchange System and Services) is an interactive data network for bidding, offering and trading commodity futures and options offered by the NYMEX. The system currently trades crude oil, heating oil, gasoline, natural gas, and propane. Realizing the limitations of its new financial products, particularly electric power, NYMEX has decided to move all these new products to an electronic environment totally and dispense will floor trading. On March 3, 2000, its five electricity contracts were moved to ACCESS with the capability to trade the contracts 23 hours per day. Its coal contract will not be launched on the floor but in an electronic format as well later this year. A Middle East Crude oil futures contract for the Asian oil markets will be launched on its electronic trading platform during 2000. And now the exchange has announced the development of its own Internet trading platform—eNymex—which will incorporate OTC contracts.

- **California Power Exchange (CalPX)**—The California Power Exchange (CalPX) was mandated by law by the California Public Utility Commission as part of the deregulation process there. The CalPX is a public company. It became active on March 31, 1998 and primarily uses a power pool, or poolco model. Energy pricing is not actually known until electricity is dispatched. It functions as a central power exchange using a computer-based auction process providing a market clearing price (MCP) for qualified buyers and sellers or electricity within the state of California. As part of the deregulation agreement, the three large investor-owned utilities (accounting for 80% of all electricity used in California) must participate in the power pool and must buy and sell their electricity through the California PX. No other state, region or country has adopted the California model of complexity nor will ever.

- **Automated Power Exchange (APX)**—The Automated Power Exchange is an Internet-based electricity trading exchange, which first became operational in the California market on March 30, 1998 when California opened its power market. It is privately held and backed by Bechtel Enterprises. The APX is a certified power scheduling coordinator between the California Independent System Operator (ISO) and electricity generators, retailers and end-users active in the California markets. The APX opened exchange markets in Illinois and New York during 1999, and will open an exchange in the UK and Japan during 2000.

- **Altra Energy Technologies**—Altra Energy Technologies serves over 4,000 users and is clearly the leader in electronic energy trading. It is not an Internet platform yet, but will be moving to the Internet by June 2000. Their platform trades gas and electricity online.

  On October 14, 1999, Altra Energy Technologies, Inc., Amerex Power Ltd., and Prebon Energy Inc. announced the launch of Altrade(tm) Power, the real-time, online electronic power trading system, provided by Altra Streamline, L.L.C. The launch represents the first business-to-business e-commerce site to unite new online trading exchange technology with the strength and reach of the traditional voice broker market. It is hybrid solution gives customers the ability to execute orders electronically without having to forgo the benefits of voice brokering. It is an interim solution for pure electronic trading, but presently has attained the high ground in this increasing competitive electricity trading marketplace that lacks futures liquidity and price transparency. It is the first system launched but not the only one. Recently, BP Amoco, American Electric Power and Koch Industries bought small equity positions in Altra.

- **Houston Street Exchange**—Houston Street Exchange also known as HoustonStreet.com launched an on-line electricity trading floor for wholesale energy traders on July 8, 1999 tailored for the Internet. The system initially covered the NEPOOL (New England), PJM (Pennsylvania, New Jersey and Maryland) and New York State markets. It is a separate subsidiary of BayCorp Holdings, which owns physical generation assets in New England. Houston Street is positioning itself as an online broker for US wholesale power markets, posting bid and offers as well as completed trades. It is a relatively unknown entity that has spent a considerable amount of money on creating an advertising buzz and continues to do so. Although limited gas and oil applications have been added during first quarter 2000 and although Equiva and Williams have made investments in Houston Street, the company has had limited success in attracting liquidity to its platform. The exchange will enter both Europe and Asia.

- **RedMeteor.com—**RedMeteor.com is a web-based, neutral online transaction exchange for crude oil, natural gas, natural gas liquids and electricity for the North American markets initially launched on April 3, 2000. The system will eventually be expanded to the emerging markets of emissions, weather and bandwidth trading. The system has the capability for trading physical, financial and options products. The system is designed to be have customized information for traders including NYMEX streaming, 24 hour accessibility, weather and news information, and most importantly access to real-time bids and offers. At the present time, it is the only transaction-based online trading Website for oil, gas and power. RedMeteor.com does not take title to any product transacted on its site. The breakthrough of this system is the ability to conduct multi-commodity trading on the web. Relating to each commodity, crude oil trading can either be a physical or financial transaction on RedMeteor.com. Electric power will trade at the major delivery points in the US. Natural gas will trade at the major hubs. Natural gas liquids are either used as a hedge with crude oil or separately by producers and end-users.

- **IntercontinentalExchange (ICX)—**On March 21, 2000, a group of leading US and European financial institutions and some of the world's leading energy companies announced the launch of IntercontinentalExchange which will be their new Internet-based electronic marketplace to trade over-the-counter price swaps and options for energy, metals and other commodity products. Initially, the venture will begin trading crude, heating oil, gasoline, jet fuel, silver and gold in the third quarter of 2000. Contracts offered on these products will include forwards, options on forwards, contracts for differences and swaps. The trading platform will expand to include global natural gas, electric power, and a variety of base metals. The companies involved in this venture include BP Amoco, Shell, Totalfina Elf Group, Societe

General, Goldman Sachs, Morgan Stanley Dean Witter and Deutsche Bank. These are mostly major European players in energy and metals trading. This exchange will have no memberships, no dues or fees beyond those incurred to process the transactions.

The seven founding firms will provide initial market liquidity of $20 million apiece and have committed to a minimum level of participation hopefully ensuring liquidity in that exchange. The trading platform is effectively a principal-to-principal business but is directly competitive to the NYMEX. Today, the vast majority of OTC energy and metals trading is conducted by telephone trading networks. The notional value of the OTC commodity contracts was more than $1.8 trillion in 1999 and is still growing robustly. Moving to an electronic platform should improve operational efficiencies and reduce costs. The ICX will also attempt to create a paperless back office that would permit market players to clear and settle OTC contracts. This electronic OTC market could prove tough competition for established futures exchanges.

- **Petroleum Electronic Pricing Exchange (PEPEX)**—The world of oil market tendering is about to radically change by a new Internet-based company. Everyday, about 50 million barrels of crude oil and refined petroleum products are transacted through market tenders. This archaic process suffers from antiquated technology, language differences and widely different time zones. Oil is a 24/7 commodity that moves globally through tankers and pipelines all the time. That system will be changed forever with the launch of the Petroleum Exchange Pricing Exchange (pepex.net). PEPEX was launched on March 25, 2000 and is the first web-based platform to move the tendering process for crude oil and petroleum products to the web. PEPEX allows the tender initiator to submit tender offers with quality specifications, quantity, and delivery dates to tender receivers on the Internet. PEPEX charges as transaction fee per barrel to the winner of the tender when a deal is consummated. The system will provide electronic documentation and product specifications, multiple language capability and real-time availability. The movement of the tendering process to the web is a major breakthrough in the world of oil trading where traditionally telephone and telex were the modes of operations in an ultra secretive mode of conducting business.

  PEPEX has also made an agreement with Empresa Colombiana de

Petroleos (Ecopetrol), Colombia's national oil company, to a two-month deployment of various trades to its client base of about 100 companies on May 1, 2000. Similar agreements being negotiated with several other state-owned oil companies, principals, governments, and large consumers. It is a major breakthrough in moving state-owned oil companies to trade and hedge their price risk.

- **Enron Online**—On October 26, 1999, Enron launched a global Internet-based transaction system for wholesale energy and other commodities. Enron OnLine is the company's Internet trading unit and will allow participants to view and trade these commodities in real time with Enron over the Internet with no commissions. Enron acts as market maker. The transaction system went live on November 29, 1999 for North American natural gas with other products to follow during 2000. Enron's customers will have access to several hundred traded products around the world. The company will publish wholesale real-time prices for electric power, natural gas, coal, weather, liquids, petrochemicals, pulp and paper, emission credits and other commodities in North America, Europe and Asia. It is expected that Bandwidth trading will be added during the summer of 2000. The backbone of Enron Online trading is Enron's own fiber optic network, which exceeds 12,000 miles of fiber routes and is expected to surpass 18,000 miles by 2001.

  Enron Online does not match buyers and sellers. Instead, customers will directly transact with Enron as the principal. The system is targeted for retail customers and wholesale and will eliminate the need for Enron to spend so much capital on marketing. It is expanding into bandwidth and emissions trading, and will also manage credit risk. Since its launch, the system has already captured $9 billion of business through early march 2000. But the key factor is that Enron has moved its entire commodity portfolio to the web, and is in effect cannabalizing its existing business model.

- **Enermetrix**—Enermetrix.com Exchange was launched on August 23, 1999, and has refocused their efforts as an energy procurement exchange that uses the web to match buyers and sellers of energy and energy products. The company feels that is provides a missing link between wholesale power trading and retail energy service markets, and can provide significant energy savings for its customers through the use of the

Internet. In effect, it is an Internet-based auction service similar to e-Bay but targeted to energy buyers and sellers.

SCANA, South Carolina's electric utility, introduced its own Internet auction in late October that was targeted for energy suppliers and end-users in the commercial and industrial sector. SCANA has created SCANA Online with Enermetrix which will be an electricity and gas auction marketplace. Qualified suppliers enter bids to supply and deliver energy at the best price for SCANA customers which post their energy needs on the Internet. The best bid then is accepted and a real-time transaction for pre-qualified suppliers takes place. This new process eliminates the Request for Proposal (RFP) process. The system is for all commercial customers nationwide not exclusively for SCANA customers. A similar arrangement was contracted between Public Service Electric & Gas and Enermetrix in February 2000.

NORTH AMERICAN EMISSIONS TRADING

Since market mechanisms are expected to be employed to decrease the cost of national emissions targets, there are a number of companies either readying or launching electronic emissions trading platforms. This section of the report will examine the status of Bloomberg, Natsource and Cantor Fitzgerald's systems. This effort will accelerate beyond the current sulfur dioxide (SO_2) and nitrous oxide (NO_X) efforts as greenhouse gas (GHG) emissions (including carbon dioxide) become more important to meeting corporate emissions reduction schedules. Thus, the next emissions trading market that is emerging concerns CO_2 emissions reductions which will most likely trade on an electronic platform.

- **Bloomberg PowerMatch**—Bloomberg's PowerMatch system allows the anonymous trading of OTC electricity products for wholesale power in North America as well as trades emissions allowances for SO_2 and NO_X with approved counterparties. This electronic broking system allows multiple bids and offers for the same trade to be visible over the Bloomberg system.

- **Natsource's Quote System**—OTC broker Natsource has launched a web-based SO_2 and NO_X quote system, which provides traders with real-time updates and indicates when a trade is made. Natsource also offers a live quote screen system for coal trading on the OTC markets. The system is accessed through Moneyline system, a financial service provider on the Internet. They provide the e-commerce interface for Natsource. Natsource's online platform supplements its traditional voice broking system. Thus, it is another hybrid electronic trading platform.

- **Cantor Fitzgerald**—The Cantor Fitzgerald brokerage house developed the concept of electronic trade matching for the highly liquid US Treasury cash market. Their system uses an interactive matching technology that trades on a time priority basis with goal of providing more liquidity to the market. The Cantor Exchange (CX) launched in 1998 to develop an electronic exchange for the trading of US treasuries, and its a niche in electronic trading of treasuries will provide its launch platform into emissions trading. The system is now called eSpeed. Cantor Fitzgerald's environmental brokerage services include the buying and selling of SO2 allowances, OTC NOX allowances, greenhouse gas reduction and other environmental credits. During April 2000, Cantor has also extended its platform into gas and electricity trading through eSpeed, which has Dynegy and Williams as partners.

THE EUROPEAN MARKET

The European energy market provides a potentially lucrative platform for electronic trading. Two factors support this view: first, unlike the US market, Europe does not have a traditional underlying infrastructure of exchange platforms; and second, growth in European Internet usage is on an exponential growth curve.

At the end of 1999 Europe was limited to three electronic energy platforms: London's International Petroleum Exchange (IPE) which trades its natural gas futures electronically on its Electronic Trading System (ETS); Nord Pool which trades electricity forwards and futures for the Scandinavian market on the electronic PowerCLICK platform; and the Amsterdam Power Exchange (APX) which trades the day ahead electricity market on an Internet-based platform.

During 2000, the number of known European electronic platforms launching will increase by a factor of three. And it is more than likely that by the end of the year the number of electronic energy platforms either directly developed for the European market, or those that have European applications will have grown significantly. The driver for this is the increase in Internet-usage and web enabled companies. In the UK the government has initiated a policy to increase Internet usage from under 25% (usage per capita) in 1999 to 100% by 2004. Similar initiatives within the other EU member countries are also in progress.

The structure of European energy trading has undergone a number of changes, which have yet to be incorporated by trading platforms. Market consolidation, as a result of electricity liberalization, has increased both within EU member states and cross-border. The implementation of the EU gas directive on competition (August 2000) will further increase the rate of consolidation and convergence within the European energy sector.

The consolidation of utility operations will directly feed into ancillary service markets such as trading platforms. Conventional wisdom suggests that an expansion of trading platforms will be surplus to the needs of the consolidating underlying utility sector. The European market can take a more objective view of market requirements by considering the developments in the more mature US market. In the gas and electricity markets the end game is to provide a Europe-wide platform with regional trading hubs to meet traders requirements as cross-border trade develops. These platforms will provide spot, forward, futures and over the counter products as dictated by the underlying market requirements. Clearly market dynamics will determine the eventual winners and losers.

There has been an initiative towards a pan-European trading platform with a letter of intent signed in October 1999 between Nord Pool and APX, to develop a European Electricity Futures Exchange to be located in Amsterdam. The aim of the co-operation was to set up a reliable futures exchange quickly and was based on APX and Nord Pool sharing the same views on international market development with both exchanges sharing the belief that the establishment of a primary electricity spot market price is a necessary basis for trading electricity futures contracts. The rationale was that in such a European Electricity Futures Exchange the two companies would combine Nord Pool's experience and the initiative and market support on the European mainland from APX. However, the letter of intent was not progressed, but the view expressed by both exchanges that a pan-European platform is the right way forward remains. The driver in this respect is likely to be Germany based on its market size and its fully liberalized electricity market.

In March 2000, Frankfurt's European Energy Exchange (EEX) and the Leipzig Power Exchange (LPX) entered into talks aimed at a merger before the scheduled launch of either platform. While both platforms will offer different contracts they will compete for the same market. Given that the EEX has more support from the German market it is likely that LPX would assume the position of a secondary market. It therefore makes sense for both platforms to merge, providing a robust platform for the German market.

The Amsterdam Power Exchange (APX) is following its own agenda to diversify its trading operations in Europe. It concluded a deal with German transmission operating company VEW in March 2000 to launch a German hub in April 2000 for the country's northern grid and was seeking a similar venture in the southern grid.

The interplay and strategic initiatives of the main European platforms suggests that, like the underlying utility market, the future shape of the trading infrastructure is uncertain. What is certain though is that the consolidating market will not support a plethora of platforms.

As we have previously discussed, in tandem with competition among utilities there will be similar competition among the trading platforms. The question to ask is how many electronic platforms will be required to afford an effective trading system for the European electricity market?

From a market infrastructure perspective, Europe can be sub-divided either by transmission grid networks or by regional demographics. Using transmission grid networks, the European market can be divided into three regions: the UK, Scandinavia (which constitutes the NORDEL system) and the remainder of the EU15 (the UCPTE transmission system). This market sub-division would require three exchanges of which the Scandinavian market is already accounted for with Nord Pool. Likewise the UK market will be traded out of London. But the real choice is for the UCPTE system. Most observers believe the natural location for a mainland European market will be Germany.

We believe that the future European trading hubs will be located in Oslo (Nord Pool), London and Frankfurt (EEX). Within this structure the three trading hubs would be electronically linked, providing access to the remaining European market from any hub. While the UK hub would be a standalone platform, i.e. there would be no 'spokes' to other regional or 'subhubs' within the UK, the other hubs would have spokes to other regional markets. Scandinavia would be a self-contained market with Frankfurt consisting of spokes into Benelux, Iberian market, Italy and possible Central Europe (i.e. Warsaw).

In order to achieve this vision, the market will undergo a transition structure to achieve it. Such a transition structure would incorporate five European hubs with Benelux and the Iberian market joining the three hubs identified above. APX is the obvious platform for the Benelux region while the Valencia-based FC&M commodity exchange is currently assessing the potential for an electricity exchange.

As market competition increases the opportunities afforded by electronic trading will significantly increase. The issue to be addressed when considering this market's trading development is whether trading will be facilitated within a 'united states of Europe' structure or through individually linked European regions.

As with competition in the utility sector, a number of electronic trading platforms are expected to emerge over the next couple of years—both OTC and exchange based platforms. Consolidation of utilities will also be reflected in the future make up of trading platforms, which tends to support either a centralized trading platform, or trading platforms electronically linked.

We believe the vision expressed by the Amsterdam and Frankfurt exchanges for a 'European' exchange will not be realized. Unlike other energy commodities, electricity is essentially a regional market and has to be traded as such. Therefore the extent of a European market will be determined by the geographic boundaries of the transmission infrastructure.

As mentioned earlier, this leads to a future structure of three European trading exchanges, or hubs—located in London, Oslo and Frankfurt—with each of the three exchanges can be linked electronically. The genesis of this trading infrastructure is already apparent. The only unknown is whether Frankfurt's EEX grows through strategic alliances with the other mainland Europe exchanges (i.e. APX) or adopts a merger and acquisition approach. This realization of an interlinked system of regional trading platforms is only achieved through electronic and Internet-based technology.

ASIAN PACIFIC AND AUSTRALASIAN MARKETS

The Asia Pacific and Australasian energy trading markets are less developed than those in North America and Europe concerning energy derivatives trading; however, it seems likely that many of the ideas on new energy derivatives contracts such as emissions and weather as well as the electronic trading platforms can evolve rapidly in these markets. Some of the reasons for these imminent changes are the movements toward market deregulation and liberalization bringing more market risks. Recently. Enron, Texaco and Sithe Energie have announced plans or opened offices in Japan to trade Japanese electric power markets and buy IPP plant capacity. The reason why is that Japan is the second largest power market in the world after the US and has a need to drive down costs there.

But the key market driver is the easy and rapid dispersion of electronic energy platforms from one continent to another due to the rapid acceleration of technological change. For example, only two years few Japanese business people used electronic mail. Now, almost all businesses not only use email but have web sites as well. This chapter focuses on recent market developments in Australia, New Zealand, Singapore and Japan. For a more in depth reading on energy derivative market developments, one can review **Asia Pacific Energy Derivatives** (December 1997), a Financial Times Energy management report written by Peter Fusaro, one of the authors of this chapter.

- **SwapNet Ltd.**—SwapNet is a electronic oil trading system targeted for Asian and European oil markets. The thinly traded Asian energy derivatives markets should benefit from the liquidity provided by this e-Commerce offering. The Internet-based system is geared to provide screen-based trading for swaps in oil products, crude, inter-month spreads and crack spreads, which could undermine the traditional telephone-based brokers. On September 14, 1999, a joint venture agreement has been entered into by Garban-Intercapital, the world's largest OTC derivatives broker. The joint venture will see Garban-Intercapital offering electronic swaps trading in certain commodity derivatives using the SwapNet trading system in the Asian and European energy markets. It will enable third parties to execute oil swaps contracts using the Internet.

  It can be operated on a purely electronic basis, or on a hybrid basis utilizing the SwapNet hybrid broker interface. With this latter approach, users have the option of authorizing brokers to enter orders electronically on their behalf. Garban-Intercapital will provide a hybrid broker function, similar to Altra's Altrade, offering market information to traders, and entering/withdrawing orders into SwapNet trading screens as required. The trading terminal interface allows traders to view, and act upon, real-time bids and offers and provides daily summaries of trading activity. The best bid and offer in the market is always displayed on the system and bids and offers from counterparties who are not on a trader's.

- **Australia**—On November 16, 1999, the Sydney Futures Exchange (SFE) moved all its contracts to an electronic platform, using its SYCOM IV trading platform, which is a Windows NT upgrade of the SYCOM III system. They will close their trading floor. According to the SFE, this enables it to retain third generation tried and tested functionality, while achieving greater flexibility for enhancement and scalability for future growth. The new platform will have multiple exchange capability, allowing members to trade all SFE and New Zealand Futures and Options Exchange contracts from the same trader workstation, whenever each market is open, implying trading on a near 24-hour basis. An Automated Order Entry Interface (AOEI) will allow members to pass orders electronically into the trading system form their order management, order routing or proprietary trading system.

  The system uses the following algorithm process to execute a custom order: First, system attempts to match a custom order against the book of existing custom orders; second, if the order is either unable to trade in the custom market or is partially filled, the system will attempt to trade the order or partial order in the outright market as a market Fill or Kill (FOK) order, maintaining the correct ratio assignment. This means if it cannot trade immediately it will be retained as a standing order in the custom market. Each time a custom order is hit in the custom market the system will attempt to trade the remaining volume in the outright market as a Fill or Kill (FOK) order. The remaining volume will be retained in the custom market. A custom order can never trade directly against an existing spread order, however, it can trade against real or implied prices in the outright market.

- **New Zealand**—The division of state-owned Electricity Corporation of New Zealand (ECNZ) into three competing state-owned companies—Genesis Power, Meridian Power and Mighty River Power, and the launch of the country's retail power market in April 1999, have both increased the potential for competition and encouraged the development of more trading platforms. In September 1999, a new electronically traded forward contract was launched by Wellington-based commodity broker Ord Minnet. In response, NZFOE is considering the launch of a similar electronic peak contract, to augment its base load futures contract, but has to resolve Y2K issues first. It also has to mitigate for the Pool settlement process—whereby prices are settled eight days after transaction—which has limited trading liquidity. The new Huntly forward contract requires all participants to sign ISDA agreements to allow for pricing variations following the settlement period. The consensus is that OTC electronic trading is the only way to increase trade liquidity in the New Zealand marketplace, and it should lead to a number of new platforms to compliment the new restructured market.

- **Singapore**—The development of the Singapore Power Pool during 1999 was the first step to a competitive market in this region. A number of companies are looking at the opportunities for IPP development with the financing securitized by selling into the Pool. This should open up trading opportunities although no developments have yet been proposed.

  An obvious platform will be Singapore Exchange (SGE), which has the NYMEX ACCESS electronic trading system, and a mutual offset system for the International Petroleum Exchange (IPE) Brent crude oil contract. SGE, in line with other financial and commodity exchanges will be moving fully electronic early in the new millennium. Its current energy contract portfolio is limited to IPE Brent and fuel oil, although the addition of an electricity contract will be a welcome extension of its portfolio in light of power market developments in the country.

- **Emissions**—The Sydney Futures Exchange is developing the world's first exchange traded emissions contract. The concept of the emissions trading scheme would be to allocate carbon sequestration credits whereby a buyer can offset CO2 emissions. The emissions trading system would be based on a permit for the holder to emit a fixed quantity of greenhouse gas. The SFE would develop the underlying carbon sequestration credit. The denomination unit would be for one ton of CO2. The initial effort would focus on Australia but the concept can be rolled out throughout the Asia Pacific region on carbon credit trading. The exchange would trade these credits electronically.

Another market seriously looking at emissions trading is the Japanese energy market. It is envisioned that a financial market for Japanese oil, gas, utility and industrial companies will be very viable in this heavily energy intensive market. This will probably on an electronic platform.

SOFTWARE TRADING SYSTEMS

The underlying structure of the energy markets makes them extremely conducive for the deployment of risk management software systems. The first wave of risk management software in the mid 1990s came from either the foreign exchange or oil platforms. Today, however, there is a need for an integrated systems approach for energy risk management. The need is to integrate and capture all data affecting the energy trading business in real time. That means the fuel inputs of oil, gas, and coal as well as the generation output of electricity but also taking into account the volumetric risk of weather and the government-mandated risk of emissions. This platform also must be flexible and robust enough to incorporate both wholesale and retail market risk management. Clearly, today's energy risk management platforms must evolve to the higher level of integrating both the physical and financial markets. Electronic trading just adds more complexity to the energy risk equation.

In electricity trading, additional complexity is offered by transmissions and scheduling needs, retail market aggregation, and power outages. Other changes underway that will influence the software solution are the future need to meet gas market pricing on an hourly basis. Weather risk and emissions trading just all add more complexity to the energy stew. Clearly, traditional risk management software solutions are not adequate to meet the emerging needs of the marketplace. The rapid rate of change in the energy markets has created a vast amount of applications handling different aspects of the business. Consequently, Information Technology is a very important function in controlling risk. Electronic trading moves all the energy commodities to a real time environment where speed and accuracy are both important. The extreme volatility of oil, gas and power markets brings that much more complexity to handling risk.

Indeed, it should be recognized that a 'total' end-to-end energy solution is probably unrealistic given the complexity and volatility of the energy markets. By its own definition, an end-to-end solution would theoretically make the job of the trader virtually redundant. Further, an end-to-end solution for one company may not work for another company. This realization tends to enforce the concept of application system provision whereby a systems company works with a utility to develop the trading infrastructure and support that meets the utilities specific requirements. Broadly speaking, the configuration of an end-to-end solution would require four key platforms: front-to-back office, e-commerce, consulting, and data integration.

Obviously any software system has to be compatible with the trading system employed by the trader. As trading moves into the electronic era systems have to interface with it. Current systems tend to be categorized as electronic communication networks (ECN's), but the rapid advancements in technology development and application will likely make ECN's virtually redundant in the next few years. ECN's will then be replaced by electronic negotiation networks (ENN's), which will provide a stronger and more proactive interface with the underlying trading system.

Further down the road, convergence of many different products and services aside from electricity or natural gas will enable consolidators to price all products under one contract. Internet bill payment is a natural use of this technology solution. Commodity markets in fact promise razor thin margins and the need to offer new goods and service to customer segments. Moreover, competitive market forces should reduce costs over postal and financial services that are used today. New entrants into the energy marketplace will further drive change in the traditional utility environment. E billing is one application of the technology.

The E-Business model for the electric utility business is just now evolving but the core concept is the ability to allow transactions for the business either in wholesale energy trading or in retail services for customers. Utilities are starting to recognize that the technology imperative becomes a key market driver for not only reducing customer service costs, but also a means to retain and attract customers. It improves the quality of the customer service. Incidentally, Internet back office applications like billing and customer care are becoming much more central to the energy business.

Today, some of the key barriers to electronic electric and gas bills are the lack of industry standards since the marketplace is still developing as well as the need to create an "electronic bond" with customers. It extends the reach of the utility and should improve efficiencies in utility operations. It is also bringing with it new competitors who have a different view of the industry and use different business models. Some of these efforts will fail, as a shake out in the industry is inevitable. But it is interesting that Internet information parallels energy flow and the liberalization effort now underway. It is another change agent that fundamentally changes the utility business. Further penetration of personal computers for residential as well as commercial and industrial customers will make aggregation efforts easier to become successful. In the future, the Internet will become the standard for all utility transactions, and the value of this transaction-oriented data will become more valuable.

ANCILLARY MARKETS

Newer commodity markets that impact on the energy industry are also appearing. The trading of weather, emissions, water and bandwidth are developing as long gas and power markets throughout the world. It seems the trading solution is taking root globally. These markets may or not be conducive to electronic trading depending on their state of development i.e. are they liquid enough. This section investigates their origin and where they are heading.

- **Weather Trading—**The emerging financial markets for weather trading are another area for the application of electronic trading. While the OTC weather derivatives market has consummated about 3,000 trades in the US and a handful of trades in Europe since 1997, weather derivatives providers will not imminently use an electronic trading platform. That said the latest market development is a weather futures contract. The Chicago Mercantile Exchange (CME) has prepared both a standardized weather futures and options contract that will be traded electronically in order to hedge temperature related risks based on heating and cooling degree day (the measure of temperature fluctuation above and below 65 degrees Farenheit). The CME new weather derivatives contracts trades for 12 consecutive calendar months, be accessible 24 hours using the

Globex2 electronic trading system. It was launched on September 22, 1999, but is failing to attract liquidity.

Weather protection programs offer electric and gas utilities the ability to hedge their weather exposure primarily since they have at least 20 years of data, which can be weather normalized. Thus, they can predict a linear relationship between gains and losses based on favorable and adverse weather conditions. Each degree-day change affects earnings by a quantifiable amount. Derivatives are weather exposure offer the potential to stabilize volume-related earnings or provide earnings for a minimum number of heating or cooling degree-days for contracted periods of time. Their complexity since they are a cross commodity hedge of fuel usage and power consumption will require more extensive modeling of annualized weather data before they can approach a standardized product. At present, they are a customized, one-off market, and most likely at will remain an OTC product or reinsurance product.

- **Bandwidth Trading**—Telecommunications network capacity would seem to be another natural market for commoditization due to its similarities with electricity trading. Bandwidth, which is the capacity, to transmit voice and data over telecommunications networks has the same characteristics as power trading in that it can not be stored, it has off peak and peak pricing regimes, and in effect, it is a real time commodity. Bandwidth also has regional demand, capacity spreads, seasonal demand and pooling points. Current market drivers in bandwidth trading are Enron Communications, Arbinet, Band-x, Bandwidth Markets.com, Bandwidth.com, AIG Telecom, GTX, USCOMEX, and Interion.

  From the energy markets, current market leaders are Enron, Koch, El Paso and Morgan Stanley have already created bandwidth trading desks in order to make a market in this newest commodity. Enron Communications is currently developing a fungible product tied to a delivery mechanism with a master agreement. A third party will oversee performance and quality of service. This is the next step to the full commoditization of bandwidth. It seems likely that Enron's offering will be web-based and electronic for wholesale telecommunications markets.

  Enron Communication's initial offering is two city pairs in the US: one service between New York and Los Angeles targeting Fortune 500 firms on T-1 lines, and the second will be an Internet Protocol between

Washington, D.C. and San Francisco/San Jose targeting telecommunications firms and Internet Service Providers. These configurations will serve as benchmarks for the most liquid telecommunications path in North America. They will also serve as launching points to connect the European and Asian markets.

The thought is that first movers in these markets may be hard to displace once that they have established a strong market position and brand. Early market entrants can determine a large part of how that market is shaped and establish a strong, global position. This is the strategy that Enron is now pursuing as it branches out of the energy patch into bandwidth trading at quite a rapid pace.

ENERGY E-COMMERCE

The next generation of e-commerce is beginning to emerge with the use of more seamless technology. Electronic Data Interchange (EDI) and other standards are beginning to emerge which are better encrypted and more secure. But the reality is that today, the energy electronic commerce solution is focused on the building of an infrastructure rather than exploiting more powerful network applications. In a sense, they are first and second generation technologies. Once the networks are more established and robust, even more competitive solutions and applications will emerge. Business to business e-commerce in energy will be forced to move to real time with next hour gas markets following electric power markets. Aggregators will provide more bundled services. And a true multicommodity warehouse of oil, gas, coal, power, emissions, weather and bandwidth will be available in the trading equation as a one-stop shop.

While established energy commodity exchanges fear erosion of their market franchise due to new electronic competitors, fragmentation of the market will be the immediate impact before the market consolidation period occurs. Competition will force the existing exchanges to alter their traditional way of doing business but probably cannot move them fast enough to meet the new competitive floorless challenge.

ELECTRONIC REGULATION

In a world without walls, global exchanges will be the shapers of the rules, standards and technologies. New electronic exchanges were not envisioned under the regulatory structures of open outcry trading floors. While traditional risks of mishandling of accounts and floor trading market manipulation will recede, new types of regulatory oversight will be needed. Record keeping requirements, for pit trading will become obsolete, but electronic audit trails will be needed to be maintained.

Because many of these exchanges are being created in the US, US regulatory agencies and structures will predominate initially. Market surveillance from watchdog agencies such as the US Commodity Futures Trading Commission and Securities and Exchange Commission will change as well. In effect, new regulatory structures will have to evolve since they have not been envisioned as of yet. Of course self-regulation, which now applies to the OTC derivatives market by the money center banks, may become the standard as electronic energy commerce matures.

Turning to the European Union, Europe is struggling with the same problem and is trying to pass legislation to govern e-commerce on the Internet across Europe. Since individual countries don't wish to delegate this authority to Brussels at the present time since it is an evolving situation and may have to be decided on a country by country basis each with its own set of laws and regulatory structure.

Down the road, barriers to e-commerce will be removed as the authority to regulate the Internet is borderless and will eventually shift to the broadest possible governmental authority. The economic imperatives of globalization and the Internet will eventually force state and federal regulators to cede power to the larger national or international authority, whether that is the US Government or the European Union.

CONCLUSIONS

Rapid growth of electronic trading is forcing fundamental, structural changes in the energy markets and in the energy industry. The model of global energy trading is being irrevocably changed. Better transaction data, more price transparency, reduced trading fees, and access to better information will create more liquidity but lower margins. Volume will surge, and newer players will be engaging in the business of energy risk management and energy trading.

It is predicted that the spread of the Internet and electronic commerce will give rise to price destruction on manufactured goods and fundamentally change the manufacturing industry. E-Commerce is already becoming the main distribution channel for the energy industry. The change rate is accelerating as energy trading takes hold throughout the industry. It is only the beginning of this fundamental change process.

While today Internet technologies are still prone to problems regarding reliability, speed and performance, the transformation into a medium that is fast, reliable, and convenient is rapidly emerging. Already hand held wireless devices for cellular phones and notebook computers are under commercial development and will use Wireless Applications Protocol (WAP). This change will bring seamless access to the Internet in the so-called web to wireless (W2W) marketplace. The impact on Internet energy trading will be instantaneous access in real time from anywhere in the world. The movement toward broad band technologies with text, voice, video, and graphics will widen applications even more and move past current Internet gridlock. DSL and cable modems will move more data, that is, financial transactions; thus adding the technological capability to enhance market liquidity. Moreover, speech recognition and translation technologies will be more finely developed which will further globalize Internet-based trading. These new speech recognition algorithms will improve the interface with the network creating the virtual global trading floor. Some energy market players are in fact waiting for greater technological developments before they launch their electronic trading platforms. They will use the technologically advanced edge to gain market share.

Looking at other financial markets today, the aggressive move of the Deutsche Borse to trade stocks, options and futures in the US, and NASDAQ's push into the European markets signal that global stock exchanges are not that far off in the future. The globalization of electronic energy trading may be only a vision today, but the reality is that global trading platforms can be launched and functional very quickly. This technological advantage may be the key to development a global electronic emissions trading market since carbon reduction for global warming mitigation requires a global solution under the Kyoto Protocols. In effect, electronic commodity exchanges can either be an extension of existing futures exchanges or new rivals.

The Internet has been called a near perfect took for capital markets and offers other advantages such as the cheaper alternative to physical markets. The Internet also allows the real time dissemination of financial information including pricing, spreads, and yields on a computer screen.

Obviously, the new technologies harnessed by the Internet challenge "open outcry" exchanges as electronic communications networks (ECNs) are even cheaper than established electronic exchanges such as NASDAQ. It should not be forgotten that over five years ago the very high telecommunications cost was a major impediment to greater market penetrations of Williams Chalkboard oil trading system, the first electronic energy trading system.

Electronic energy trading may be a double edge sword. It may lead to more trading liquidity with more individual investors (the so-called doctors and dentists who trade the NYMEX), but it could lead to higher price volatility since active day traders try to exploit tiny price discrepancies in the market. This trend is already in evidence in US stock trading as a "volatility influence" exists. For the energy complex, which are the most volatility commodities ever created, it probably means even more volatility fueled by day traders. This phenomenon is already in evidence and influenced by NYMEX floor traders who trade for their own account on a daily basis.

As established markets consolidate and demutualize in response to the new technologically advanced competitors, the role of existing exchanges changes to that of listed companies. In effect, the NYMEX and IPE become commodities. As the NYMEX is the oil and gas futures exchange, and continues to grow every year in those two commodities, private investors and companies can buy a share of NYMEX. This is significant since the oil and gas contracts are robust and have grown demonstrated pronounced growth during 1999.

The overall impact of electronic energy trading system is that traditional floor exchanges are fighting for survival in the wake of technological change and global financial integration. They must adapt or be superceded by the next generation of technology. These new electronic exchanges are thus perfectly positioned for the emerging markets of electricity, emissions, weather and bandwidth trading since they can be constructed quickly and at minimal costs. More refinement and technological improvement will come, as there is recognition that gas markets in the US particularly must trade on a next hour basis to match the power trading requirements. Real-time will really be in real-time in the future with 24 hour markets everyday of the year.

Electronic trading is an infinite market since the developments in technological advances is only time dependent. The current electronic systems are only the beginning. The rate of advancement in electronic trading technology will make current systems redundant in the next few years. The challenge to improve the systems on offer is immense and will increase as trading demands increase.

CHAPTER 10

Energy Indexation: Analyzing the Scope of Electricity Price Indexes

By Antoine Eustache, Ph.D.

Since the publication of the Dow Jones California Oregon Border Electricity Price Index (COB Index) in 1995 there has been a proliferation of power market indexes in the US and in Europe. Not only have these indexes played a central role in the creation of the power markets on both continents, they have fueled the emergence of an over-the-counter financial market that could some day surpass the physical power market in size. Given the global trend toward liberalization in recent years can this trend toward indexing expand into newly emerging markets such as bandwidth, emissions trading, etc? Undoubtedly some of these markets will experience major price volatility and, as a result, will see substantial demand for derivative products. Can these indexes be made robust enough to serve as benchmarks for settling exchange-traded futures, even in exceptional cases where physical delivery may present major problems? The methodological approach to creating power market indexes described below may prove relevant in various emerging commodities markets.

ELECTRICITY PRICE INDEXES

The major impetus behind the commoditization of electric energy began in the late 1970s and has been spreading worldwide. Although, in most cases, the electricity industry has made a successful transition to the new market environment, there is much about electricity that makes a commodity trading market difficult to establish. Except in markets dominated by ample hydro supply, for instance, electric energy is practically non-storable. This means electricity must be consumed as it is produced. This lack of storability makes electricity a highly volatile product. Even prior to deregulation, the marginal cost of electric power routinely reflected unpredictable

FIGURE 10.1

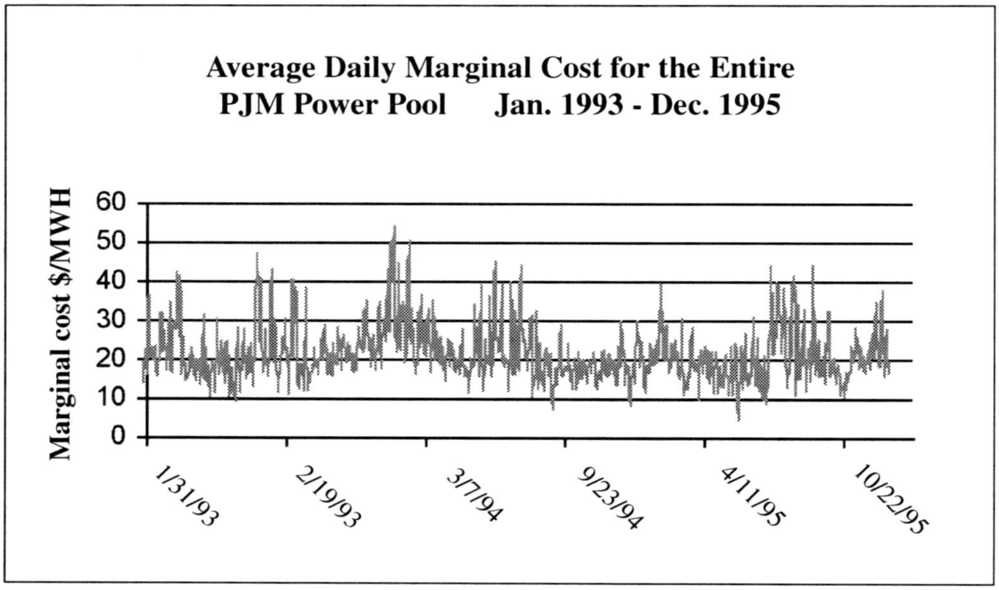

spikes during the summer and winter months. This tendency to become volatile is illustrated in Figure 10.1.

This inherent volatility became apparent as the new market place began to value power in terms of prices. It became more pronounced as trading practices shifted from long-term contracts spanning over many years to transactions that could be consummated on the spot.

Transportation is another problem area. For instance, given two transmission lines running next to each other, power will tend to flow through the line that offers the least resistance and any imbalance at any point of the transmission network can cause the entire system to fail. This and various other transmission issues tend to exacerbate the fear, especially in markets still adapting to the new competitive environment, that the commoditization of electric energy could cause a threat to the security of supply. For this reason many of these markets tended to promote the formation of various pooling organizations that not only controlled the central dispatch of power plants, but also maintained a very controlled pricing environment. Such pooling arrangements all suffer from the perception that prices are determined through rigid mathematical formulae instead of by the sheer dynamism of a truly competitive market.

Nevertheless, the over-the-counter markets have done well in most countries where the legislature does not prohibit bilateral trading. In the United States, the number of marketers buying and selling electric energy in the bilateral market increased from less than 10 in 1993 to over 500 by the end of 1999. As of September 1999, these power marketers moved over half of all market-based wholesale electricity traded. In Scandinavia where liberalization has been highly successful, hundreds of brokers and power marketers coexist side by side with Nord Pool, the Nordic power exchange. In Germany, where power marketers and brokers began trading in March 1999 in the newly liberalized power market of Europe, the trend indicates a highly successful spot market is underway.

To a large extent this success is due to a high level of price transparency embedded in the creation of various price indexes. During the early phase of market development the various efforts made along the traditional price gathering method proved inadequate for power traders. Even the most innovative approaches that introduced some form of electronic bulletin boards, which enable traders to post bids and offers did not win market confidence.

This occurred for two main reasons. First because of the complexities of the power business it was not clear to the new entrants what electricity products to trade. For instance while the electrons in a megawatt-hour of power may be the same throughout the day the premium associated with a megawatt-hour at midnight is less than that paid for the same unit at midday. Second, the perception that some power producers could not be trusted to be forthcoming with reliable information forced the market to seek a more robust means of price gathering.

Mindful of the importance reliable pricing signals play in market development investor-owned utilities, municipal utilities, power marketers, and various financial institutions agreed to collaborate in the creation of electricity price indexes first in the United States, and more recently in Europe. The creation of these indexes called for the definition of a set of products, the choice of market hubs, and an acceptable data gathering methodology. The markets also saw the need for a neutral arbiter who could not only assist in developing these indexes but also guarantee its widespread development and educate the public of its implications for the emerging market.

SETTING UP THE MARKET

In the Western United States, where the Western System Power Pool (WSPP) had set the framework for the creation of a spot market, it was easy to define a set of products that could be traded easily: day-ahead on-peak firm energy, day-ahead off peak firm energy, and so forth. In Continental Europe, where there was no central organization like the WSPP promoting the trading of electric energy, defining a set of products was more challenging. The issues became much simpler when two leading utilities in Switzerland agreed to cooperate on the development of the Swiss Electricity Price Index (SWEP) in 1998. A year later, the decision by PreussenElektra, Germany's second largest utility, to support an index on its power grid, the Central European Power Index (CEPI), removed the final obstacles to defining standard products most traders could agree on.

In all cases firm power was defined as energy that must be delivered once sold except in case of force majeur. To be considered as firm these contracts had to be backed by generating capacity. Non-firm energy was a type of contract that gave buyers and sellers some flexibility in how they chose to deliver the energy. These contracts could be cancelled at any time for any reason once adequate notice had been given.

Due to structural similarities between natural gas and electricity, power executives were quick to learn from the experiences of the natural gas industry. Having successfully been deregulated nearly a decade earlier the US gas market was a perfect example of a competitive market. Market centers emerged around the numerous interconnections of the gas pipelines. These centers, especially those with many interconnections, were called hubs of which the most famous is the Henry Hub in Southern Louisiana. Two of the major characteristics that made these gas hubs successful were its capability to bring large number of buyers and sellers together and the ease with which these agents transferred gas title from one owner to another. The large number of power marketeers that entered the electricity business were quick to promote this concept as an ideal solution to the complex issue of defining delivery points in the emerging power markets. To facilitate the creation of the first power market place in the United States, the market selected the California/Oregon Border (COB) as a hub. In reality the COB was a point of interconnection on three high voltage transmission lines linking the Pacific Northwest with the Pacific Southwest. The continued demand for power at that point and the ease with which buyers and sellers could transfer ownership of power made the COB one of the few points on the west cost with the characteristics of major market hub. Similar hubs have been designated in Europe.

The methodological approach to gathering data played a crucial role in the development of power indexes. Two of the central issues were the perception that the sheer size of some power suppliers could not only lead to data manipulation but also to off-centered indexes. On the latter issue it soon became clear that no matter how well designed, an index could never remove the structural imbalances of a market. This was understood to be not only true for the commodities markets including power, but also for the equities markets as well. However, it was possible to introduce safeguard mechanisms that would at least facilitate the tracking of manipulation to the source and, if necessary, take measures to remedy the wrongdoing. Two such mechanisms were a) the recognition of Dow Jones & Company as a neutral arbiter to help design and calculate the indexes, and b) the execution of a legal agreement between each data provider and Dow Jones. One of the key provisions of this document was the agreement by each data provider to submit to occasional audits by a professional auditor at Dow Jones' sole cost. Although none of the providers would agree on a financial penalty in case there was evidence of a pattern of deception the knowledge that a provider could be publicly dropped from the index was deemed a strong enough deterrent. The market also agreed that a volume-weighted-average formula would accurately depict the realities of the power industry.

Having agreed on a methodology and a calculation method, the data-collection process was straightforward. Each day, by a time agreeable to all parties, Dow Jones collects prices and volume of daily energy transactions. This includes prices and amounts of energy sold. The information is transferred electronically and is stored in a permanent file for auditing purposes. Because of the confidential nature of the data it is processed in a highly secured environment by staff members whose main task consists of guaranteeing that the Index is accurate and reliable. Data that deviate substantially from the mean are investigated and, in the absence of a reasonable explanation for the deviation, are removed from the pool.

How well do electricity indexes track the power market? Although the electricity market is still evolving and may not have yet reached the level of maturity of the equities markets, the evidence suggests that electricity indexes provide a reliable picture of a market well on its way to becoming information efficient. This is apparent when the Dow Jones power indexes are compared to other time-tested benchmarks. In assessing the reliability of the power indexes, the distribution of their daily returns is compared to that of the S&P 500 (Figure 10.3) and a set of randomly generated and uniformly distributed numbers (Figure 10.2). As would be expected the random data show an almost linear pattern with a Pearson's correlation coefficient of 0.9954.

FIGURE 10.2

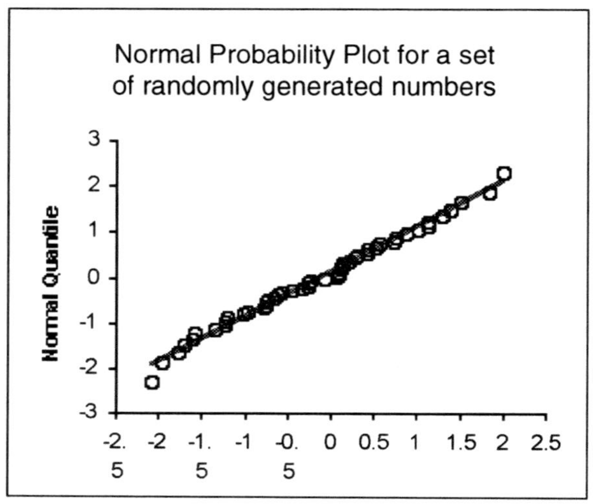

FIGURE 10.3

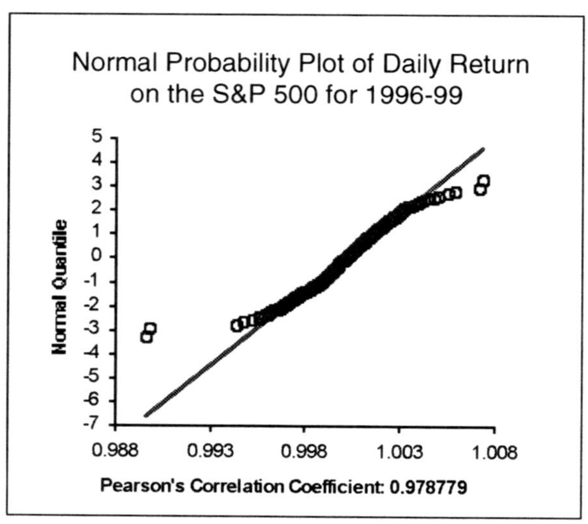

FIGURE 10.4

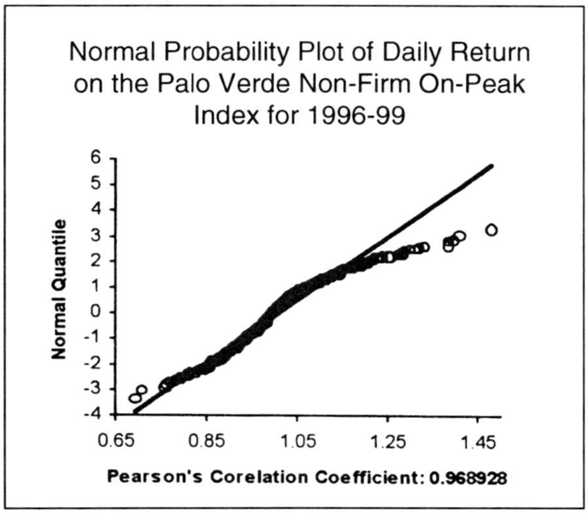

FIGURE 10.5

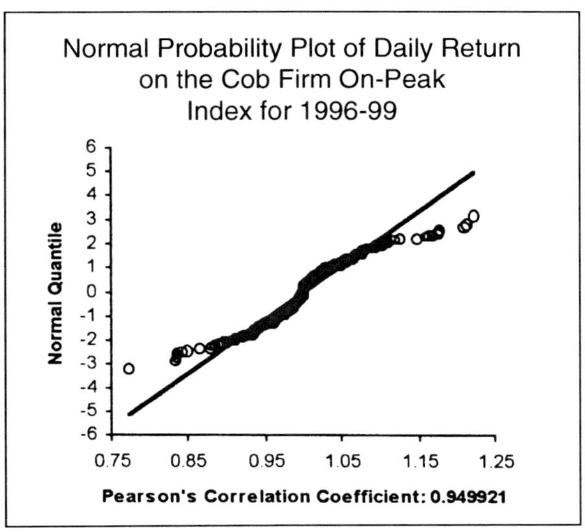

However, even without accounting for factors such as seasonality or the mean-reverting property of power prices, it is worth noting the high correlation coefficients, 0.97 for the Palo Verde Index, and 0.95 for the COB Index, compared to a 0.98 correlation coefficient for the S&P 500. This suggests that power indexes in their current states are highly reliable. They also indicate that despite the fact US electricity market is still going through an evolutionary phase, it is relatively efficient when compared to the more mature equities markets.

GETTING TO CASH EXCHANGES

Financial instruments such as swaps and options are not new to the electricity industry. Swaps are private agreements between two parties to exchange cash flows in the future based on a prearranged formula. Options are financial instruments that give the buyer the right to purchase or sell a security at a predetermined price-the strike price. The growth of swaps and options was long stifled by the lack of transparent and reliable prices. Electricity price indexes have removed that impediment by providing electricity traders with barometers that enable them to gauge the dynamism of the cash market. This makes it much easier to use swaps, options — and eventually futures contracts — to manage risk in the electricity market.

At that point, the buyer or seller of electricity is dealing in exchanges of cash, and may, through a broker, choose to negotiate a swaps deal with any credit-worthy institution willing to guarantee him a revenue stream over the period he feels exposed. It could be a commercial bank, an investment bank, a trading house — or any other large or sophisticated small investment-management firm.

There have been various arguments on the need for and the suitability of price indexes for settling electricity contracts. These arguments are in part fueled by the lack of interest in the existing futures contracts traded on the NYMEX and the CBOT. From the early days of deregulation, many financial and power experts believed that financially settled electricity futures contracts would be more viable than physically delivered contracts. This view was ignored by the various exchanges competing to be first in launching successful contracts in the power industry. Now that there seems to be a clearer consensus on the direction of the cash market, many experts are wondering whether it would make sense to develop indexes that could serve as benchmarks for settling electricity futures.

It is indeed possible to envision a scenario where electricity price indexes serve as the underlying benchmarks for settling electricity futures. The issue of firm power delivery is at the core of the debate on the viability of delivered futures contracts in the power industry. New York Mercantile Exchange futures contracts are traded on a firm delivery basis. However, if in the old days of the industry firm delivery implied that physical assets such as plant capacity and transmission access, backed firm power transactions, the modern definition of firmness is couched more in financial terms. For instance, in today's power nomenclature, "liquidating damages" is the term most often referred to when power traders seek to guarantee the terms of the deal. This means that buyers can be financially compensated when a party fails to act on a firm promise to deliver on a power transaction whether as a result of an act of God or simple negligence form the seller's side. By providing a financial definition for firm delivery traders seem to tacitly agree that a liquid power market cannot be adequately served if agreements are strictly tied to physical assets. It is therefore fitting to assume that futures contracts based on the modern definition of firmness will gain in popularity in the wholesale market. Since financially settled futures contracts call for strong benchmark prices, it would be appropriate to assume that reliable electricity price indexes would be a part of a solution involving such settlement.

As for the current offering of power indexes in the market, despite their widespread application in settling over-the-counter swaps and options and their statistical validity they may need to be more broadly accepted by the marketplace before they can be used for settling futures contracts. For instance, the Dow Jones power indexes are considered the preferred benchmarks for settling over-the counter swaps and options in the United States and in Europe. Many US states and recently one Canadian province have recommended that utilities use these indexes as benchmarks for pricing certain industrial contracts. However, as the power market evolves, Internet trading platforms are capturing a great portion of the deals and the trend is increasing. The brokerage community is also playing an active role in matching buyers and sellers of electric energy. Given this wide representation these indexes may need to capture some of these activities as well in order to give a more complete snapshot of the trading activity.

RELEVANCE TO OTHER MARKETS

Despite its seemingly unique complexities electricity shares a number of similarities with some of the markets emerging the world over. This is especially true for natural gas, coal, and telecom bandwidth.

- **Natural Gas**—Over the next several years the European continent will witness the emergence of a very active natural gas market. In the United States where the wholesale gas market has been successfully deregulated, retail competition will soon become a way of life. Undoubtedly, most of the new entrants in this new market will be institutions with vast experience in the electricity business. Because of the strong interdependence between natural gas and electricity both industries will share a common set of values and business practices.

  As more and more consumers look for new ways to save on energy expenditures, price will become extremely important not only in the wholesale markets, but at the retail level as well. In addition energy

FIGURE 10.6

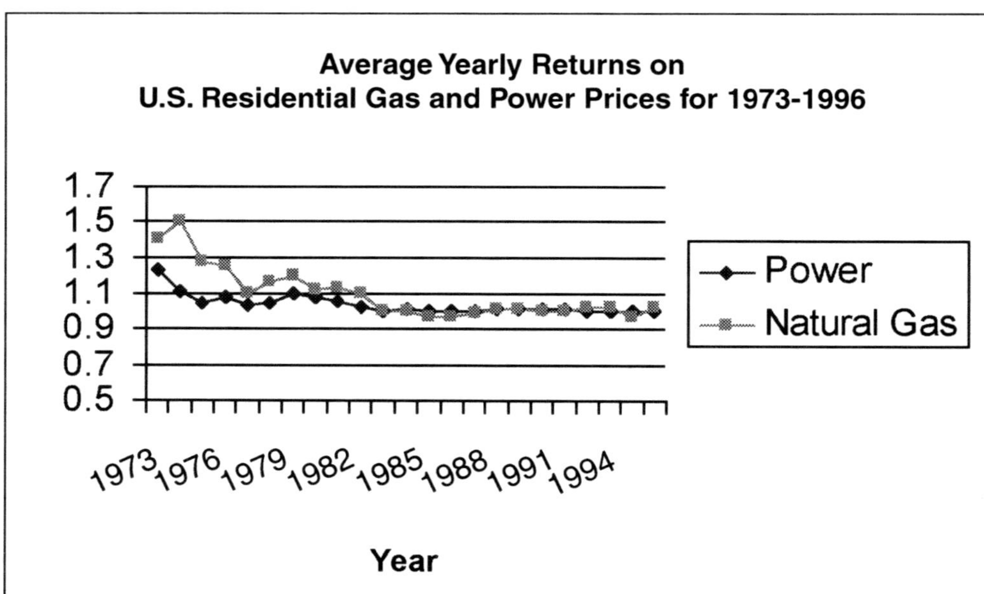

intensive businesses will be actively involved in the daily spot market for gas and power. For many of these consumers price risk management will play a central part in daily operation. As a result these businesses are expected to become active users of benchmark indexes that enable them to tie the price of gas or electric power to some financial derivatives. Because natural gas is also used as an input fuel in electricity generation, electric utilities will also pay close attention to retail as well as wholesale gas prices.

In some cases electric utilities will maintain a balanced portfolio of financial risk management products that include gas as well as electricity derivatives. Given the realities of these two sectors, it is reasonable to assume that demand for natural and power price indexes will rise over the next few years.

- **Coal**—Coal accounts for over 50% of the basic input used in electricity production both in the United States and in Europe. Until recently most utilities purchased coal based on long-terms contracts spanning many years into the future. However, since the late 1990s, there has been a growing movement toward commoditizing coal on both continents. In the United States, the New York Mercantile Exchange has been a leading proponent of several coal futures contracts. Various utilities are buying coal on a short-term basis. Others have gone as far as using a method called tolling to hedge coal and electricity price risks. For instance in a tolling arrangement, a coal supplier or a power marketer may pay a power plant operator a fee to convert coal into electricity when the market value of power is greater than that of coal.

  At the center of these sorts of transactions price remains the major determining factor on whether or not the deals will take place. As the coal market become more competitive and tolling practices gain in popularity, price indexes that allow buyers and sellers to make a one-to-one comparison between a unit of coal and a unit of power will become extremely valuable, as will the basis provided for swaps, options and futures.

- **Telecom bandwidth**—The emerging bandwidth market is another area where the experiences of the electricity market could prove relevant. Over the past three years, the global economy has experienced a phenomenal growth in electronic commerce. In the U.S. alone business-to-business e-commerce was estimated at nearly 140 billion dollars in 1999 and analysts predict this market will grow at a rate of 41 percent a year to over 500 billion dollars by 20031. Residential usage of the Internet has seen a phenomenal growth ranging from a few million users in 1994 to over 100 million in 1998 and experts believe one billion people around the world may be connected to the Internet by 2005. This convergence onto the Internet of electronic commerce and residential demand for information technology has led to increased demand for broadband telecom. The demand for broadband telecom is engendering a commodity market, which analysts predict may be worth over 400 billion US dollars by 2005 and 1.5 trillion dollars by 2010. The importance of bandwidth as a commodity takes on a special meaning for power executives. First almost every major electric utility has a large infrastructure of fiber optic network. As demand for telecom products increases and market pressure unveils the volatility of bandwidth prices electric utilities will be highly pressed to protect their bottom lines. As a result demand for risk management products will surge. However, since financial risks cannot be managed without reliable benchmark prices demand for bandwidth indexes will also increase. Because of the influence of power marketers such as Enron in this newly emerging market and the tendency to introduce the same basic trading concepts that have worked for electricity and natural gas the future telecom indexes will not be unlike electricity indexes in design. There, too, the opportunity for swaps, options and futures markets will emerge.

TOMORROW: THE WORLD?

As the experience in Europe and North America suggests, electricity price indexes are powerful instruments that have not only facilitated the creation the power market in both continents they also serve as powerful benchmarks that help simplify the risk-management practices in the power industry. This simplicity is at the core of the recent explosion in the growth of electricity derivatives in both the US and Continental Europe. As the market evolves, these indexes will not only become the benchmarks of choice in over-the-counter markets, they may also play a central role in promoting the growth of exchange-traded electricity futures. We have barely even mentioned emissions trading, another promising derivatives venue. This experience can be applied to other goods, perhaps including goods that do not exist today, as they become commoditized and widely traded.

---

[1] U.S. Department of Commerce, (1999) The Emerging Digital Economy. Washington, D.C.

CHAPTER 11

Managing Risk in Retail Markets

By George H. Campbell

EXECUTIVE SUMMARY

We performed a recent study that showed every retailer in the competitive US electric market is losing money. Is the lack of profitability due to poor performance from the traditional risk manager? The answer is no. Most electric marketing efforts are doomed from the start due to poor risk management entering the market. This chapter not only explores what traditional risk management means in retail markets but also identifies the risk(s) that must be managed before entering the market.

There are five principal areas of risks that must be managed in the retail market. Some are self-evident and are not thought of as traditional risk. First it is important to identify and manage the type and magnitude of potential losses that can be incurred prior to establishing a traditional risk management program.

The five principal risks in the new retail market are in the following categories:

- Geographic market risk
- Market segmentation risk
- Pricing and contractual risk
- Energy and transmission procurement risk
- Settlements, billing and accounts receivable risk.

FIGURE 11.1

Figure 11.1 shows how the sequence of risk should be managed in the market. It starts with assessing the risk in the geographic market and finally proceeds to the settlements, billing and accounts receivable risk. Choosing geographic markets and selecting target segments in these markets is the first step. If these are managed first, then the traditional risk management function may be able to deliver the value needed to make the business profitable.

GEOGRAPHIC MARKET RISK

If profitable geographic markets are not chosen, there is very little hope of earning a profit, sometimes for more than a decade. The two drivers to selecting the right market are the regulators' and legislators' attitudes toward competitive markets and the actions of other retailers in these markets. The attitude of regulators and legislators is the main driver to profitability in retail access markets. As seen in California and other states, it is not the high price of retail rates or the magnitude of stranded costs that makes a good market, but the regulators' and legislators' attitudes. Will they encourage marketers to profitably provide power priced lower than the utility bundled tariff?

During state restructuring proceedings, there was significant testimony given by many marketers predicting there would be significant savings because markets were more efficient and marketers could purchase energy better than utilities. Actually, that has not proven to be true. The savings in retail access programs have resulted from the following:

- Legislators allowing utilities to refinance their debt for stranded assets through Government guarantees;

- Regulators not allowing utilities to recover 100% of their stranded costs; and

- Social and environmental program restructuring, as seen with PURPA, Demand Side Management (DSM) and other programs, which were raising electric costs for all consumers.

The states where regulators and legislators want to encourage competitive retail markets allow the majority of these savings to be passed along to customers, rather than give a general rate reduction for all customers. These rate reduction options have always been available for regulators to lower overall rates, but they are often integrated into the legislators' and regulators' restructuring decisions so that they can claim "everybody will have lower rates." In California, New York, and New England, the regulators and legislators, for the most part, took the benefits from securitization of debt and non-recovery of stranded costs and lowered residential and small commercial rates. In Pennsylvania and New Jersey, the regulators took some of these benefits and enabled the generation shopping credits to be set high enough to allow the marketers to potentially earn a profit.

The other market characteristic affecting profitability in geographic markets is the number of retail marketers that are subsidiaries of regulated utilities and have as their marketing goals acquired a high percentage of market share. Even if a potential profit margin exists for the retail marketer, if your competitors are bidding retail prices less than wholesale to obtain market share, then there will be no profits in the immediate future for any retailer. Unregulated retail marketers of utilities often believe that their future business is acquiring retail customers. Market share, not short-term profitability, becomes their driving goal. Risk is usually significantly shorter in duration as earnings and cash flow are reported to the financial community and pricing discipline takes place.

MANAGING RISK IN CHOOSING MARKET SEGMENTS

Once your company has determined it is going to enter a certain geographic market, the next risk that will have to be managed is choosing a market segment. Usually retailers have already predetermined which segments they are going to pursue in retail access markets. These decisions are based on their own core competencies as a company and/or because they have done research and believe that there is more profit in one segment versus another.

Speaking as a retailer that has looked at most markets in the United States, until the retail access rates are set, you can't determine which segments will produce enough margin over the bundled utility tariff rates to justify your cost to serve them. A sure sign of potential disaster is when you hear a retailer say that even though they have not analyzed the retail access tariffs, they are going after a certain market segment because they believe in their core competencies. The following figure shows the sequence of how risk should be managed in choosing a market segment.

FIGURE 11.2

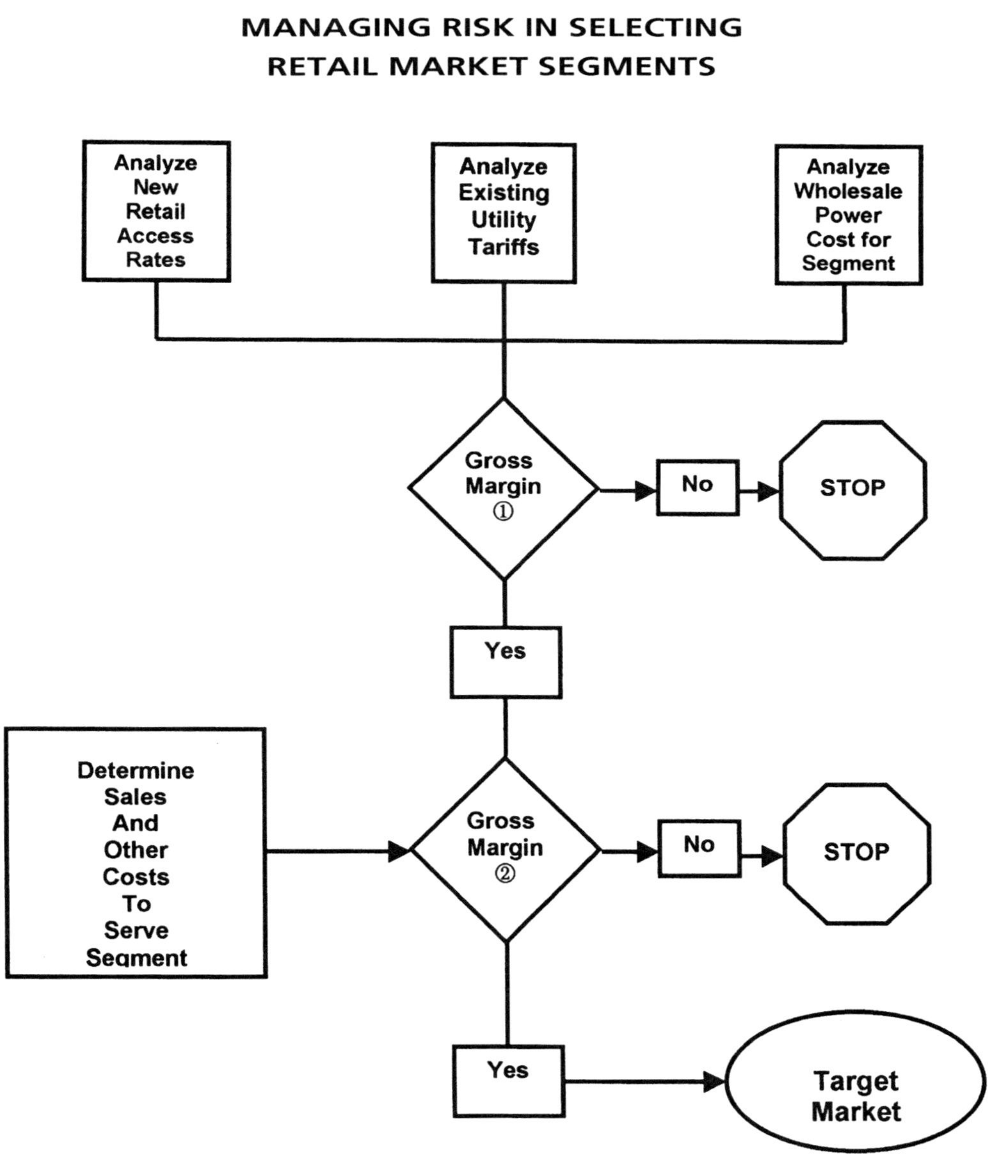

As the flowchart (Figure 11.2) shows, the first thing that has to be done is to analyze the retail access tariffs, the utility tariffs, and the wholesale power costs to determine whether an initial gross margin can be made. If a gross margin is identified, then estimates have to be made on the sales and mid- and back office costs to serve this particular segment. If there is still a margin, then these tariff segments become your potential target markets.

Intra-rate analysis also needs to be performed on retail access rates. The rate class analysis, as a whole, may not produce enough of a gross margin to make a profit. However, within existing tariffs, there may be certain types of customers with seasonal loads or certain load factors that will produce a larger gross margin. If intra-rate segments produce a satisfactory margin, then the cost to identify the customers and to sell to a smaller market niche has to be determined.

Retail marketers must also manage the retail access tariff risks and other rules and regulations after the retail access market opens. Retail access rates and rules and regulations are often reviewed and changed at regulatory commission meetings. Ongoing administration by the regulated utilities can significantly change a retailer's profitability for a certain segment. Retailers, as a group, must be willing to manage the risk in this process.

MANAGING PRICE RISK

In typical markets, developing pricing products means understanding what the consumer wants and delivering it. In retail access markets the answer is simple — getting a lower price than the bundled tariff price from the local utility. Even though it sounds simple, it can be very complicated and extremely risky depending on how the retail access rates are set. In addition to a lower price, consumers want the same or less risk than they had with their existing utility tariff.

In California, the first state with retail access in the United States, the retailer has to purchase at a lower cost than the hourly Power Exchange (PX) price. Retail access rates are set in California by taking the customer's existing tariff rate and subtracting the hourly Power Exchange price after the billing period is over. The California PX is an hourly wholesale electric energy spot market. Therefore, to beat the existing tariff price, customers must receive a discount off the hourly PX price. Because there are no forward products sold that are a "PX minus price", risk management in this market is the proverbial "crap shoot". Why would anyone sell at "PX minus price" when they can sell to the Power Exchange and get full PX prices? A risk manager can't manage this price risk because the fundamental economics in the wholesale market are such that nobody would offer this wholesale product.

In most other states, the retail access rates are firm rates and do not vary within the year as the wholesale market price changes. Price risk management essentials in these markets are setting prices based on the forward markets and managing volumetric risk.

To date, however, prudent price risk management has often been ignored due to immaturity in the electric market and the pressure to obtain market share. Retail marketers have underwritten significantly more price risk than their risk-adjusted returns will justify. However, if their competitors are willing to accept this market risk (and probably lose on the deals), and there is still pressure to obtain market share, price risk management becomes an arbitrage function. The risk manager's job will then focus on eliminating more risk than your competitors while still meeting sales targets.

Now that we have talked about the "gloom and doom" of risk management in new retail markets, let's assume that the market matures, there is not an incumbent utility offering an unbeatable bundled tariff rate, and your competitors have pricing discipline. How would a risk manager oversee price risk under these conditions?

In its simplest form, the retailer would take the forward wholesale product and design retail pricing products. In addition, load-shape and volumetric consumption risk would have to be incorporated into the price products.

Price risk management in retail electric markets is a relatively simple concept with complicated tasks to manage. Here are a few reasons why:

- A highly variable underlying spot market that varies from \$0.00/KWh to \$10.00/KWh;

- Hourly scheduling and balancing on the electric transmission system; and

- Liquid forward wholesale pricing products are designed for trading and not for retail price risk.

Because of the highly variable hourly spot market, load shape risk is significant. The following example illustrates load shape risk.

Two customers have 1,000,000 KWh per year of consumption. Both have 10% load factors. One customer is completely off-peak with a wholesale price of $0.01/KWh. The other customer has load factors at the highest priced hours and the price is $0.20/KWh. These wholesale prices are indicative of present hourly spot markets in most regions of the country. Therefore, offering a customer a flat price, without tying them to a shaped load consumption pattern, is extremely risky for a retailer.

In addition, volumetric risk has to be managed so that when forward products are bought and prices locked in for retail volumes, customers do not significantly change their overall consumption and leave the retailer exposed with un-hedged volumes. Some of this risk may be hedged with weather derivatives, but it is better to manage this risk in pricing products rather than pay for weather derivatives.

The standard wholesale electric transaction in the United States is a daily, weekly or monthly 16-hour peak product. The 16 hours are for hours beginning 700-2300, Monday through Friday on the East Coast and Monday through Saturday on the West Coast. However, the underlying spot market significantly varies from this flat price. Figure 11.3 shows a peak day in SERC and how a flat 16-hour price would be extrapolated to follow the hourly spot market. As shown, the price is significantly below the forward price for the first and last hours of the peak block. However, it is significantly above for the afternoon hours. Managing retail customers' price risks mean normalizing the flat priced forward products for the contract term. Because the underlying spot market continuously changes as the market matures, normalizing prices never results in a perfect hedge.

FIGURE 11.3

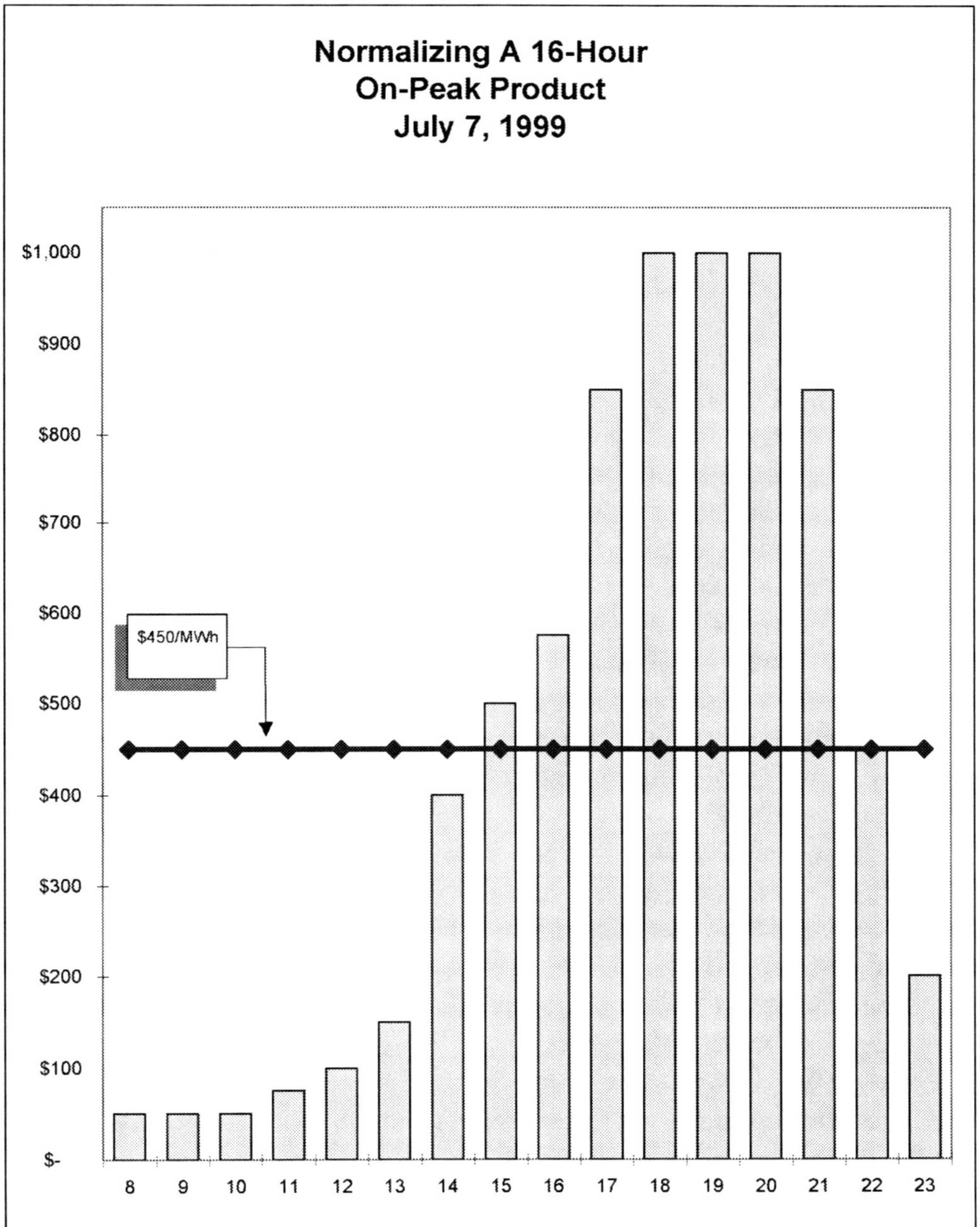
Normalizing A 16-Hour
On-Peak Product
July 7, 1999
$1,000
$900
$800
$700
$600
$500
$400
$300
$200
$100
$-
$450/MWh
8
9
10
11
12
13
14
15
16
17
18
19
20
21
22
23

There are three basic ways of managing the highly volatile hourly price risk with pricing products. They are:

- Multiple tiered Time of Use Prices (TOU);
- Very high fixed charges and energy prices; and
- Block and swing rates

Multiple tiered, four tiered, or "time of use" price structures will manage hourly price volatility, especially during the high price months. When consumers change consumption patterns, their unit energy price will vary and more closely follow the hourly wholesale spot market.

The second methodology that could be used involves charging very high fixed charges based on the actual or nominated monthly demands. These fixed charges would pay for the premium prices in the top 300 hours per year. The fixed prices would be weighted to the months where the high hourly spot prices occur in the market. There is significantly less energy price volatility in remaining hours of the year, and a one- or two-tiered TOU price would cover the load shape risk. Also, if a nominated demand charge was used it would cover most volumetric risk. Any consumption greater than the nominated demand should be priced at the wholesale spot market. If actual demands are going to be used as the basis for recovering the fixed charges, a premium must be received to cover the volumetric risk that would be un-hedged.

The third methodology is a block and swing price. The block, a fixed load shape with a take-or-pay feature specifically designed for each customer, manages both the price risk and the volumetric risk for this portion of the customer's consumption. The swing, rates based on the hourly wholesale market or the day-ahead wholesale price, manages the price and volume risk for the variable load.

These three basic pricing options have significant variations within each, depending on the retail marketer's perspective and the customer's needs. When a marketer starts to obtain significant market share, the diversity within the customer's consumption patterns produces increased hedging, as long as another retailer doesn't identify your more profitable segments and target them.

ENERGY AND TRANSMISSION PROCUREMENT RISK

The risk manager's goal is to lock in the energy and transmission requirements with financial forward products as the customer's pricing terms are committed to a contract. All or a significant portion of the pricing risk would be eliminated.

The electric market is different from most other commodity markets, especially natural gas. In the Natural Gas market, a futures contract (almost always the Henry Hub NYMEX contract) is purchased to match the customer's volumetric needs. Then an "Over the Counter" basis swap is bought to hedge the geographic basis risk for the customer within that group. This simple hedging methodology is employed for the vast majority of the U.S. gas market.

Even after the U.S. electric market matures, purchasing energy and transmission to hedge risk will be significantly more complicated than it is in the gas industry. We have discussed within this paper many of the reasons for this complexity. Additionally, the electric market is complicated further by having its generators and loads interspersed throughout the U.S. whereas the gas market has only a few major geographic areas producing natural gas. Naturally, these generators are segmented into thousands of markets with pricing differentials due to line loading and reliability constraints on the transmission system.

There has always been natural congestion in the electric power grid, but the utilities usually homogenized these costs and allocated them equally among all customers. With new retail access markets, the transmission grid is being segmented by these natural transmission constraints, with each segment having its own wholesale price.

In addition to the larger geographic markets (Cinergy, Entergy, Palo Verde etc.), the grid is further divided into zones or nodes (substations) around natural transmission constraints. In the California ISO, 22 zones with different prices have been implemented. In the Pennsylvania, New Jersey, and Maryland Power Pool, there are approximately 1,400 nodes (transmission substations). If there is no congestion, there will be a homogenous wholesale market price for the regional market. However, when there is congestion, leading to higher priced local generators operating to support reliability in a particular node or zone, the market price increases based on the cost of the "must run" local generator. The transmission system operator will require market retailers to buy local generation for any customer in a constrained node or zone. The owners of "must run" generators know that they have an extremely valuable asset and will try to get the highest market price possible—always above the general market price. When the market matures, there will be basis risk products developed for many of these nodes and zones. In the meantime, the retailer will have to absorb the risk or buy generation output within the local node or zone.

Determining whether to buy firm or non-firm transmission is another risk to consider. Many areas of the United States have no significant transmission constraints, and lower cost non-firm transmission can be purchased throughout the year. For the profitable retail marketer, understanding the risks involved in buying non-firm transmission is another area to manage.

SETTLEMENT, BILLING & ACCOUNTS RECEIVABLE RISK

Managing settlements, billing and accounts receivable risk in the retail electric market is basically the same as in most other markets. This paper will not attempt and in-depth exploration of managing these risks. There are four distinct areas in the electric market that distinguish it from most other markets. Those areas are:

1. Settlements with the transmission system operators;

2. Delivery risk from your energy supplier;

3. Measurement of profitability for your retail customers; and

4. Collections without the ability to disconnect.

Retail customers in nodes and zones can have up to 43,800 settlements per year depending on the transmission system operator's requirements. This happens because there is hourly energy cost, three generation based ancillary services, and others costs that are also procured hourly. A transmission system operator's administration staff will be interpreting procedures and policy in parallel with the implementation of the competitive retail market. Significant money is often involved when the settlement bills are sent to the marketers. A risk manager must manage transmission settlements in conjunction with the procurement function as required until the

U.S. electric market matures.

Counter-party credit risk for energy delivery is critical in electric wholesale markets. When hourly prices exceed $1,000/MWh, counterparts may not be financially able to deliver the energy. This happened in the summer of 1998. Insuring that your energy seller is financially capable of backing the sale is critical during the high-priced months. Otherwise, you may have a significant portion of your retail power supply portfolio un-hedged as many marketers discovering during the bankruptcy of the Power Corporation of America.

Risk management principles should also drive the profitability measurements for each customer or class of customers due to the complex nature of the electric market and hourly scheduling and balancing. Customer accounting systems should be implemented to insure easy profitability measurements for each customer. Because of the load consumption pattern and volumetric risk, determining profitability requires significant systems integration. At a minimum, the two main systems, scheduling and CIS, need to be integrated.

The last major area of risk that is different from most other markets is collection risk. The incumbent electric utility is able to collect receivables much more efficiently due to the ability to disconnect. If the ability to disconnect for non-payment is not shared equally in the market, the incumbent utility or another marketer that has this capability will have a cost advantage.

SUMMARY

The existing retail electric markets have significant risks outside of the traditional risk management functions. If these risks are not managed, then the risk manager may not be able to bring value to this market as they have traditionally done. The input of the risk manager must be brought into the decision making function before the markets are established. Retail marketers will not usually be able to manage the risk assessment process in the early stages of business development.

Managing price risk in the present retail electric market is an arbitrage function. If a prudent price manager managed all of the risks associated with the new retail markets, they would probably advise their company's management not to enter any of today's markets. However, if that were the situation, there would be no risk managers in the retail access markets.

Therefore, a risk manager's job is to insure that the company is more profitable than its competitors and to understand the company's ability to stay competitive in this market. This will insure that as the electric market matures, the company will be left standing to compete in the market.

CHAPTER 12

End-to-End Energy Solutions

By Jim Banks

The energy industry has rarely seen the kind of changes that have shaped it so dramatically in the last few years. Traditional methods if buying and selling power have been eroded worldwide and a new structure has rapidly evolved into the space created by changes in regulations governing the world's markets. Alongside regulatory changes in favor of deregulated energy markets the major factor to shape the face of the energy industry as it enters the new millennium has been information technology. The advances in trading, settlement and data management technology which have given the traditional financial markets a major overhaul since the early 80's are now spreading into the acres of fresh and fertile ground the energy sector offers.

The boom energy market turnover has given rise to an unprecedented need for precision and efficiency to be second to none, as the room for competitive maneuver between rival power marketing and generation companies continues to shrink. The implication for IT companies is that in order to retain a share of a growing market they must offer a range of products, which covers all the eventual needs of their customers. The answer is a comprehensive, end-to-end system that can handle all transaction and data management needs a client may have across all energy markets globally.

According to John van Rensburg of Alstom Esca, 'end-to-end' means taking any kind of trade from the cradle to the grave. There are no half measures when it comes to attaining such a distant goal, but energy IT firms are already well down the path to it. David Samuels of Enron's EnergyDesk.com believes a more integrated, sophisticated solution is required than the current patchwork of systems currently available within most companies. "The solution must simultaneously improve efficiency, lower transaction and operational costs and facilitate more complex trades. The result is a single approach to trading and risk management from trade execution to settlement for both physical and financial energy markets."

THE INTEGRATION EQUATION

There are four main areas which energy technology companies need to integrate into one system in order to complete an end-to-end solution, and they each exhibit a different set of market characteristics which serve to make that integration problematic. Perhaps the easiest area to address at present is the trading portfolio that a company must run in order to balance its input and output, as well as hedging its exposure to the fluctuations of the market price of electricity. Solution providers like Caminus' Zai*Net have an advantage over some of their rivals in this area thanks to the expertise they have gained in the financial markets. The same paradigm used for trading capital and futures markets can be applied, with some alterations, to the energy markets, particularly given the changes that have made the operation of these various markets more homogenous. As derivatives become the main tool for hedging and trading electricity, so the systems that trade other kinds of financial derivatives are able to incorporate the energy products. For energy futures, the systems must simply learn to handle contracts with a new set of specifications.

While this area may be simpler from the IT side, however, it still represents a major cultural shift for many companies on the trading side, which may have been involved in the energy industry for many years and are familiar with a very different way of doing business. This cultural shift is an issue for IT providers, as their customers need systems that can be easily accessed, easily operated and easily understood, while maintaining vital standards of security, speed and efficiency. Utilities, for example, who have seen the wholesale market restructure dramatically, do not want to be left behind in the wake of companies with more experience of trading in a deregulated market. To complicate matters further, more and more companies are asking for scheduling and despatch operations to be included in the same area, and that they be handled by the same systems which cover the trading functionality.

OUTSTANDING LONG-TERM CONTRACTS

The second factor which systems developers must include in a comprehensive system's functionality is the capacity to monitor and analyze remaining long-term contracts for supply that must now be evaluated in the light of new market conditions. These contracts pose problems as they are not tradable, and there is anyway no liquidity in a market for 15-year deals. While these contracts may be rare, their bearing on a company's overall financial profile must still be analyzed and incorporated into any enterprise-wide assessment of a firm's risk and credit exposure. The problematic rigidity of these contracts is compounded by their often complex structure. Such long-term deals often have many physical aspects written in, and tend to exhibit embedded options, all of which makes them difficult to analyze. Addressing long term contracts, Caminus' Nick Perry maintains that these contracts are not in the same operational area as the trading portfolio, which makes them so difficult to organize and analyze, but they have to be accommodated.

RETAIL PORTFOLIO

For many energy companies the retail sector is the core of their business, and as such the systems employed to manage it are of crucial importance. This area is becoming increasingly contract-based, but these contracts are of a very different type to those in the trading and wholesale portfolios. The difference is due to the fundamental fact that the consumption of energy is determined at will by the customer, and is therefore not readily quantifiable. The management of the sector is radically different from that of the trading portfolio and centers more on customer relations and billing issues, which in their turn pose a major analytical challenge to IT developers.

FIXED ASSETS

The final business sector that needs to be taken into account is that which covers the appreciation and depreciation of a firm's fixed assets, mainly the infrastructure responsible for generating, processing and supplying power. It is important that a firm be able to factor in the impact of changes in and costs from its infrastructure and incorporate the data into firm-wide analyses.

Again the organization and management of this sector is substantially different from others. One solution has been to model the operation of some infrastructure on the behavior of financial contracts. The operation of a power station has, for instance, been modeled as an option on the spread between the fuel price (required to power the station) and the price of the electricity it generates. This, however, is a very difficult and complex task, and may not give an accurate representation of the impact of increased power generation on the company's books.

The desire to model as much of a company's operations in the style customary to the capital markets is partly due to the fact that managers of the various business sectors come to like the disciplines exhibited by the trading team. They like that trading culture factors in variables such as value-at-risk, daily reports and marking to market, and the prediction is that these disciplines will be adopted across all sectors in all companies, in order to extend the same tight discipline to all parts of the business.

The integration of these diverse business sectors is one of the barriers to the successful implementation of a truly end-to-end solution. However, the clear definition of these areas means that the goal for systems providers is in sight, and the companies involved in the development of end-to-end systems can clearly focus on the end result. As yet no single system provider has successfully addressed all these problems, but there is great momentum behind bringing all aspects of energy operations into one single analytical system.

According to Perry, some market players think that it won't be possible to bring the retail side into the analytical model, but there are a lot of companies, and Enron is a good example, which would certainly want to make this work. On the software side it is possible to come up with analytical tools, which can capture data as is. From here we can work on the theory that any determinate commercial obligation can be deconstructed into forwards, futures or options, and cash. So we can work on breaking down various elements like contracts into these different instruments, but the physical risk is a major, serious indeterminates, as is credit. Bespoke analysis is needed and this can only come with sheer brainpower and analytical skill. Perry believes the market has to deal with the messiness of reality, and factors such as power plant operation will remain a challenge to model.

INTERNET

The Internet has inevitably been a major factor in bringing together the diverse functionalities that are required for an end-to-end solution. Many recent mergers, joint ventures and acquisitions among energy IT firms have centered on the importance of Internet technology for data transmission and remote access to systems.

Excelergy Corporation, a provider of web-based customer information and transaction management technology to the deregulated retail energy market, has for instance, allied itself with Siebel Systems, a leading supplier of web-based customer relationship management (CRM) systems. The combination of services is intended to provide a comprehensive and integrated web-based customer relationship management, enrollment, and billing solution for the deregulated retail energy market. Excelergy, as a premier software partner in the Siebel Alliance Program, will add its web-based customer information and transaction software to Siebel's front office applications—another step towards offering the energy industry an end-to-end customer information management solution.

Guerry Waters of the Meta Group says the alliance between Siebel Systems and Excelergy applies leading technology to build and maintain customer relationships with extremely high levels of service, adding that it's an important step toward providing energy companies with a complete web-based customer information solution that works across multiple suppliers and markets.

Enron, a major player in the US energy markets with a capitalization of £34 billion, is also moving to exploit the Internet for the convenience of its customers as it attempts to bring more functionality on-line. Mikael Nordstrom of Enron's EnergyDesk.com believes web-based trading and risk management services are a fast becoming a necessity, not an option, for energy customers of Enron who want to do business in the liberalized energy markets. From Enron's point of view the main advantage of the Internet interface delivering risk management capability through a simple and accessible browser front end is that web-based energy risk management services are made available to all players in the market without the need for large financial investments in expensive hardware.

The importance of the Internet to companies like Enron is supported by the trend of liberalized European electricity markets requiring more innovative technological solutions. And the use of Internet functionality is crucial to this process of innovation. In the increasingly competitive energy market, distribution companies need to focus on their core business—the distribution of electricity to customers—and to reduce the time and administrative costs involved in the traditional process of power purchase. The Internet is the key to both enabling electricity to be treated like the commodity it has become through liberalization and to increasing price transparency in the energy market overall. Quickly. It is ideal for fast, convenient, easy-to-use services such as free, customized price quotes.

BIG FISH, LITTLE FISH

IT companies are well aware of the highly competitive nature of the market, and realize also that in-house system development means heavy investment of both time and money. It is the nature of the IT market that it evolves quickly, with advances in technology, such as the Internet, driving new functionality. The markets the technology must service scarcely remain static for long either, so time in particular is a luxury that systems providers can ill afford. Most firms have therefore chosen to fill many of the gaps in the end-to-end chain by spending their money on company mergers and acquisitions, as well as on development.

The many companies wishing to extend themselves into providing end-to-end solutions are not all necessarily approaching the issue from the same angle. Each has its own unique areas of expertise and experience that have developed over the years, either in energy markets or purely financial markets. Each, therefore, has different holes to plug in its suite of systems before it can claim truly end-to-end functionality.

The trend to consolidate through acquisition among IT firms is hardly surprising when one looks at the activity of firms in the energy markets they cater for. Consolidation through mergers particularly is a major feature of the deregulating electricity markets at the moment, as firms jockey for position in the newly created trading world. The major players in end-to-end system market have been showing much the same behavior as many of their customers, with similar goals in mind. Small firms with niche markets, or expertise, are being bought by the bigger players wishing to extend their services in terms of expertise, volume of customers, or geographical location. Increasing internationalization of the markets has meant a new set of issues for any IT company wishing to gain a significant market share.

A particular feature has been the desire of US IT firms, with greater experience of operating in deregulated energy markets, attempting to gain a foothold in the deregulating markets of Europe. This provides rich pickings for those firms that take the chance to take a shortcut to a system that has already been tried and tested in a deregulated environment. It is notable that the UK's daily balancing market in gas is being operated by a consortium including US vendor Altra Energy Technologies.

A powerful example of the 'hands across the pond' approach is the merger of UK energy consultancy Caminus with US financial IT firm Zai*Net. Caminus now offers the Zai*Net suite of products, which has amassed a sizeable client base in North America. Caminus' expertise in European markets and its strength in market consultation opens the door for these products to take off in new markets.

Consultancy is seen by some as vital to an end-to-end service, but for others the focus is purely on systems. Even in the consultancy area there is a boundary to be marked out between systems consulting and market consulting. Caminus' edge is that it offers market consultancy in addition to systems consulting, which the company employing expertise in the trading of the markets as well as expertise in system design. Following the deal between Caminus and Zai*Net, the new energy enterprise then went on the acquisition trail itself.

Perry says, "The company's strategy is one of build and buy. We buy if we find a gap in our cover and find a company which is suitable and consistent with our approach. All the mergers we undertake are willing mergers, and they are all private companies owned by their founders. We have never taken anyone over to take out a competitor, and we have also done a great deal of systems building."

The Caminus / Zai*Net venture is also a good example of a company which has learned its trade in the financial sector teaming up with an expert in the energy market. As with Zai*Net, so with another financial systems player—SunGard. The first step SunGard took into the energy markets was to acquire energy risk management software firm Primo Systems. With Primo Systems came the ORG group and its consultancy capability. Expertise and increased funding again combine to fill a hole in the value chain.

Scott Olley of ORG says that SunGard has a widespread presence and it needs to provide a consulting facility. "Other elements of end-to-end capability take place over several years, like straight through processing (STP).

We are big in STP in the financial markets, and we will try to leverage this into the energy markets. There is a lot of leveraging that can be done from the millions of dollars that have been invested in the banking arena."

Again, Primo Systems was a small, expert company, with a good reputation and was willingly absorbed into its larger parent. It was able to quickly expand its size and the scope of its development work. Like many acquisitions the smaller company has been largely left to operate as before, but with the benefits of increased financial support and a broader knowledge base to tap. At the time of the takeover Mark Walker, founder of Primo Systems said it was still the same company, but backed by a multi-billion dollar corporation.

The other lead player in the end-to-end market is Houston-based Altra Energy Technologies, and it is no surprise to find that it too has been buying to build. In January 1999 it bought QuickTrade to expand its customer base by around 25%. It ran the QuickTrade trading system in tandem with its own Altrade system before merging the two platforms to give it a single base from which to add functionality. At the time the company said its strategy was to integrate all of Altra's products into a single enterprise-wide, end-to-end energy solution.

With a single trading platform in operation Altra then bought troubled TransEnergy—until then a competitor in the market for energy marketing, transportation and risk management systems. "We are helping to create an energy marketplace in which transactions are consummated and scheduled via the Internet," said Altra's Paul Bourke. "By linking these transactions to integrated software solutions, the industry achieves the benefits of reduced transaction costs, improved customer service, lower error rates and improved information management."

To further enhance its web capability Altra joined a venture with two energy brokers active in serving major oil companies, gas producers, electrical utilities, power marketers, pipeline companies, institutional banks and fund managers. Prebon Energy, a division of financial and commercial broker Prebon Yamane, and OTC energy brokerage Amerex Power. The result was Altrade Power—an on-line, real-time power trading system aimed at the traditional voice trading market. Incorporating brokers active in the market was designed to allow the service to retain some of the features of the voice broker relationship. Brokers input market intelligence, and there is an element of the personal relationship maintained, but the transactions themselves are executed via the Internet. Altra describe this as "leveraging the best of both on-line electronic resources".

Then, at the end of 1999, Altra also teamed up with Excelergy, which specializes in web-based customer information and transaction management technology. The business alliance was specifically aimed at providing a complete wholesale-to-retail solution for energy by adding Altra's trading, risk management and wholesale systems to Excelergy's customer information and billing solutions. This was later supported by the acquisition of Unified Information, which added scheduling, management and settlement software for the electricity market to Altra's range of functionality.

Alstom Esca, one of the pioneers in the development of product-based, real-time control systems for the electric market and utility industries, needed to complete its suite of systems on the wholesale side of the electricity market by expanding its product line to include trading and risk management software. "Our trading and risk management software will complement and interface with our existing product line," explained Alstom Esca's John van Rensburg, pointing out that there are few trading and risk management software products currently on the market that address energy customers' long-term needs. "Until now, only a few trading and risk management software companies have had the resources to bring their products to the global marketplace. We have a history in the physical and regulatory side of the business, but the complementary products and capabilities needed to offer a complete wholesale trading solution. With these resources, we can now provide a full end-to-end solution."

Alstom, as a large international entity, saw their entry into the end-to-end market as a major marker in the development process as it brought expertise in the physical markets together with experience in the financial markets. Yet while they put themselves ahead of the pack in terms of bringing products to the global marketplace, the line from Alstom is still that the internationalization of trading leads end-to-end systems developers some murky waters.

Systems now need to handle data from a multitude of other systems across a whole posse of international markets, and the problems which stand in the way of this are all too familiar to those involved in clearing and settlement in the capital markets. The on-going work on STP is still grappling with these issues. Efforts in the area of STP have been focused on finding ways to make data homogenous, or at least easily translatable by automated means, making data capture, deal confirmation and settlement automatic and eliminating human input from the loop. Once a trade has been entered on a system it can then be executed and settled quickly and efficiently by the systems in place, without the need for a human operator to confirm or correct data.

In the absence of standardized international data protocols, which would allow all transaction details to be sent in the same format and be recognized by every trading and settlement system world-wide, there remain issues around data management and data repair. As the core principle of end-to-end is the ability of systems handling separate aspects of the retail and wholesale energy markets in a way which can be integrated across all systems, the same issues around communication, data recognition and process automation come to bear on the energy industry. Definitions of end-to-end also tend to include the notion that a system would offer global coverage across different international markets.

The two main approaches to these issues centre on establishing universal data protocols that can be recognized across markets, or on automated data recognition and translation systems which can act as an electronic Tower of Babel. ORG's Olley believes the STP concept will take time. "This is not because of the technological side, but because it will take a long time to get international participants to accept the common standard. The market needs to catch up with the technology."

THE FUTURE

At present some models might provide a suite of products which offer an 80-90% solution to the end-to-end issue, but if the current pace of technology and analysis development continues, then it will only be a matter of a few years before the current problems are dealt with. The present trend towards acquisition by consolidation allows for companies with a niche area of expertise, either in systems, consultancy or analysis, to be merged into larger entities to fill the gaps that exist in their larger suites of products. At the moment, however, the advice from the industry seems to be to equip now with a near-complete solution in order to avoid losses incurred through lack of preparedness for the prevailing market conditions.

The truth is that no system supplier can yet say that they offer a truly end-to-end system. No-one can yet incorporate all the variables in the physical and financial sides of all energy markets in all countries. However, this is more due to the nature of the market than it is to expertise in IT, so the gap will close, and the pace of integration and acquisition is not relenting yet. One systems developer posed the question: "Should you wait five years for a new theory and a complete system or should you use the 80% system?"

CHAPTER 13

An Architecture for Flexible Trading

By Michael Coleman

Energy trading is the new frontier, a vast untapped market with enormous potential. Energy is, after all, the ultimate commodity. Everybody needs to use energy whether for domestic use or for industrial process. Similar in many ways to financial markets, energy has the same ethic, intangible quality to it. It occurs, like money, in many different manifestations, coal, gas, oil and electricity. With this likeness acknowledged there are significant differences to the financial markets in other respects.

Two radical forces for change are wreaking havoc with the old established ways of procuring energy. The first is the global trend towards deregulation. The second is the revolution in information technology. The pace of change is staggering. Existing energy companies need to re-invent themselves and new energy companies are emerging from the amalgamation of elements from the financial, energy and information technology industries. This new breed of energy company will have the ability to trade in wholesale markets, to produce energy commodities such as gas and electricity as well as the ability to supply retail customers. It may well also switch energy consumption into other tradable commodities such as metals, cement, or water purification.

The game has changed radically. There was a time when power stations used to buy gas on long term fixed price contracts, used it to produce electricity as demanded by industry and the public and sold it for some margin above its entirely predictable costs. In the new energy markets, depending on minute by minute price changes, units of virtual energy will fly around cyber-space from counterpart to counterpart ending up as domestic electricity consumption, steel mill production, or conversion from gas into electricity, depending on marginal price differentials at that given moment. There may come a time soon when steel companies make their profits from energy trading and not from steel production.

The risks that energy companies face in the new trading game are orders of magnitude higher. The game could be become passing the virtual parcel of energy around until the whistle blows. In this game you don't want to be left looking for the last megawatt in the land when the price has spiked and you need to buy. The revolution in information technology makes this virtual energy-trading world a reality. Advances in the technology such as the Internet, such as the availability of cheap powerful desktop hardware and advances in software design techniques mean that the systems are now available to support a highly sophisticated electronic market place.

There are only a few stops left pushed in. Once all the stops are out the energy trading game will be in full swing. The remaining stops still need to be pushed in. Local regulations and trading practices have not been finalized or fully understood by the market players. Organizations are still reorganizing themselves to be trading orientated. Information systems are not yet fully operational. Still it is only a matter of time before these are resolved and region-by-region across the industrialized world liquid energy trading markets emerge.

The whole proposition is entirely dependent on information technology. Decision support systems and risk management systems are the key. Without such systems it would be folly to enter the game. One of the key problems though is to understand what systems are required. What should they do, how should they work and a whole host of ancillary questions.

Before answering the detailed questions and implementing systems it is worth understanding the principals. There is a set of key principals, which must be thoroughly understood and then adhered to. Once these principals are understood in an organization then it should be possible to implement systems that will support the energy trading business without making too many mistakes. The principals are quite straightforward and an architecture of system design can be built around them.

The principals are:

- Flexibility
- Adaptability
- Configurability
- Scalability

Taking each in turn as applied to the energy trading markets:

FLEXIBILITY

The problem is that no one quite knows how all of this is going to turn out. There are a number of key commodities; electricity, gas, and oil. Although other commodities play a part such as metals, for example. Nobody quite knows the degree to which the interplay between these commodities will affect the way in which markets evolve. Another issue is that the energy markets at present are really localized regional markets each with its own set of rules and regulations. Nobody knows how these rules will develop as localized markets merge and cross over.

The principal of flexibility when applied to energy trading requires architecture of design that allows for the fundamental nature of the market, which is that it is changing and it is complex. It is no good therefore to implement systems by traditional selection techniques. Traditionally the process of selecting systems is first, hire a consultant to do a "selection process", second, analyze the way business is done, third, analyze software packages available and select one that is closest to the business needs, and fourth, implement the package. What's wrong with that? It sounds very sensible.

First of all business needs today will be different to business needs in six months, which will be different to business needs in one year, and so on and so forth. An analysis of business needs is not so important as understanding the nature of change in the market place. What are the fulcrums around which change will take place? Selecting a package to fit the business needs, what's wrong with that? Well if the needs are in a state of rapid change then it is more important to ensure that any package selection made has the architecture that will support change. The package selected must be designed around the same pivotal points, the fulcrums in the market itself. The people who designed and built the package must be the sort of people you can work with, in partnership because you will need to.

The new process of package selection should therefore be as follows. Ensure that your business understands the pivotal points of change in the emerging global energy markets. Find a software package provider that also understands the same, someone whose people you can work with in partnership. Make sure that the software package has been designed using an architecture for flexible trading and implement it.

What are these pivotal points and how do you design systems that enshrine the principal of change?

The pivotal points of change are:

- **Innovation in types of trading.** Once a market becomes liquid, speculators move in, which in turn encourages more hedgers to participate. There is a tendency towards standardization of trading but at the same time a tendency towards innovative trade structures that mirror the changing needs of hedgers. As markets become more established the lines of demarcation between speculators and hedgers gets blurred. Participants in the market can be a mixture of both hedger and speculator depending on their current position.

  If the energy markets follow the same pattern as financial markets then spot trading will lead to forward trading, expanding the horizon of forward trading. Forward trading will give rise to derivative trading of cheaper and safer trading structures such as futures, swaps and options plus a never ending cross fertilization of structured deals. This is highly likely to be the way of things in the energy markets.

  When markets are new the players begin by making up the rules at the microscopic level (working within statutory regulations at the macro level). As long as the counterparts agree on the terms of their deal that's what matters. As the market becomes more established standards emerge which tends to reduce the number of variations on a theme. "Plain Vanilla swaps" emerge with standardized terms and conditions, for example. Curiously though, this standardization allows for greater innovation in the types of structured deal which traders' dream up. Structured deals end up being layer upon layer of standard deal components. It has to be this way otherwise it would be impossible to hedge structured trades.

  Information systems for decision support and risk management must be designed so that they can allow the progression from simple spot trading to a highly evolved derivative market place. This is not a question of crystal ball gazing how the market may or may not evolve. It is more a question of incorporating the fundamental building blocks into the system design so that the building blocks can be fitted together and adapted according to any unpredictable pattern that may arise over the course of time.

- **Multiple Commodities.** At the present time the individual markets in each commodity are fairly distinct and separate. The oil market is the oil market, the gas market is the gas market and the electricity market is the electricity market. This is changing and will continue to change. Once again it is not a question of predicting how it will change in any precise way. It is more a question of understanding the fundamental drivers and ensuring that systems for trading and risk management also encapsulate these principals. If the fundamental principals are enshrined in the design then it is possible to adapt to change as and when the changes occur.

  What are the fundamental principals of a multi-commodity market? The first thing is to recognize that any liquid market has certain characteristics. Supply and demand, for example, in unconstrained free markets price is the key variable. If you can estimate forward expectation of price then you can trade forward and mitigate risk. At any given moment there is a "market view" of the terrain of forward price. In a multi-commodity market the price of each commodity is correlated. Electricity is the common denominator. A series of questions need to be asked and answered at each moment—is it cheaper to convert from gas to electricity, or sell gas? Is it cheaper to buy electricity and produce steel or sell electricity? Is it advantageous to generate power in region A and swap it for power in region B and so on and so forth? Unlike financial markets the multi-commodity market has other constraints including production rates of gas, electricity network frequency, gas storage injection and retrieval rates, location of physical, transmission costs, transmission losses, price mean reversion and so on. There is a whole domain of physicality that affects the demand /supply price equation.

  Information technology systems must be designed with a number of fundamental design principals in mind. Any commodity transaction can be dissected into basic components. These basic components or building blocks are as follows:

  - Quantity of commodity (this can be real physical commodity or it can be virtual or "notional"),
  - Location (origination and delivery),
  - Date (which includes time, possibly to the nearest minute)

- Amount of money (i.e. the price of that quantity of commodity and the currency denomination).

- Counterpart with whom the transaction is made

If the information system can handle any commodity, any time period, any currency and any location, and can treat these elements consistently and simultaneously then it has the basic requirements for dealing with an unknown future. From these elemental building blocks the full range of functions can be performed. If these elements do not exist as "components" or "objects" within the system then it is likely that major work will be required to incorporate change as and when it occurs.

MULTIPLE MARKETS (REGIONS)

Energy has a schizophrenic nature. On the one hand, it is transmutable, and on the other hand it is not. On the one hand, you can convert gas into electricity, and you can do this in physical reality using a power station or you can do it virtually using a spark spread swap. A spark spread swap can have exactly the same net effect on the market or on the individual financial performance of a market player as if gas had really been used to produce electricity. Virtual power plant is available now on your desktop computer. But on the other hand, you cannot run a television set on virtual electricity. At some point reality has to be involved and this brings with it all the real physical constraints of the local market place. The physical network of wires has a real physical capacity constraint. The specific idiosyncrasy of the location where trading is taking place will partially determine the way in which trading is conducted in that location. Local governments, authorities and regulatory bodies will always have a say in the way that trading is done. This means that regardless of the global nature of trading, regardless of the merging of markets there will always remain the need to treat trading locations individually. The ways in which gas, electricity and oil are traded will always remain different.

Even though markets are merging and attempts are being made to ensure that trading arrangements are as similar as possible, just in the very nature of the way in which the different commodities are used differences will remain between commodity markets.

The selection of a software package must take into account this principal of multiple markets. If the architecture of the system did not incorporate the principal of multiple markets then major reworking will be required to incorporate a new market region or a new market sector. It may well be impossible to make such changes and the useful life of the package could be very short. Usually information systems evolve and are not really designed or engineered with this principal in mind. Typically a system is built to facilitate trading and to manage risk in one location and with a single commodity. No thought is given to the likely problems of extending to more than one location and or more than one commodity. As an after thought it becomes very difficult to incorporate such a notion.

MULTIPLE CURRENCIES

The issue of multiple-currency is really a special case of multiple markets but it is sufficiently important to warrant special mention. Trading different locations can involve trading across national boundaries. Settlements for delivery of commodity may therefore need to be made in different currencies. An easy enough problem to deal with you might think. At the point of invoice simply convert from one currency to another. Surely that is all there is to it?

Looking more deeply into the problem however, there are important issues that need to be considered. Trading in energy is changing rapidly into a real-time 24-hour market. Long term fixed price deals, which used to be the order of the day, are being replaced by a combination of short term trading and longer term forward hedging using derivative instruments. It used to be a simple matter to subtract the price of what you paid for purchasing a consignment of commodity together with any other running costs from the price at which you sold the commodity. That gives you your profit. In the new paradigm of energy trading that is no longer the case. Forward trading effectively changes that notion completely.

Profitability must be measured against the current state of forward prices in the market versus the position being held by the trader. Using current spot and forward price expectations the value of a trading position must be benchmarked, or marked-to-market against those forward price expectations. That is a more reliable indication of profit and loss. It is not desirable to wait until settlement to discover that what appeared on paper at the time of transaction to be a profit is in fact a substantial loss.

Calculating the true mark-to-market value of a position means that it is necessary to take into account the time value of money. A dollar in three months has a different value than a dollar today. The principals of discounted cash flow must be applied to calculate the mark-to-market value. If you are dealing in more than one currency it is therefore essential to know what discounted cash flow rates to apply to each currency. That means that whenever the value of a trading position is re-calculated a reference must be made to the corresponding forward interest rates in each currency. For each forward position on each forward date, the calculation must be made using its corresponding forward discount rate in the correct currency. This becomes significantly more complicated when the forward position involves derivatives trades, in particular option positions.

The point here is to illustrate that multi-currency issues are interwoven deeply into the fabric of decision support and risk management information systems. If multi-currency was not a pre-requisite design principal as part of the overall principal of designing for multiple location trading then it is unlikely to sit comfortably as an after thought.

ADAPTABILITY

The second principal, adaptability, determines the degree to which the software package can be adapted. Ultimately any software package can be adapted to any degree because it can simply be re-written in its entirety. So the question of adaptability is more to do with how quickly can adaptations be made. How reliable will the adaptations be? Will the adaptations do what they are supposed to do? How stable will the system be after adaptation? To what degree will the adaptations produce unexpected errors to occur in other parts of the package? There are other aspects of adaptability such as who can make the adaptations. Do they need to be made by the package vendor only, or could other software developers make them? To what extent can adaptations be ring-fenced? In other words, if I add a new trading instrument, is it necessary to make changes throughout the entire system so that every function in the system recognizes the new instrument or can other parts of the system automatically recognize and cope with the new instrument without further adaptation?

These issues have plagued software engineers for the last forty years.

There have been several attempts to solve the problems of adaptation, several major fashions, with each fashion lasting for around a decade. An example is the Structured Methods fashion. The idea here was that you thoroughly analyzed every minute aspect of the business that you were building software systems for and documented your findings to the nth degree. By then following strict procedures when building the software system then this would eliminate the problems. Unfortunately, the problems did not go away and the fashion moved on. Another fashion is 4GL and software proto-types. The idea here was that if you used high level tools to define the way that a system should work then you could hand the whole process of system building over to the people who use the system with a little help from software professionals. The principle being the users of the system are the best people to design it because they know what they want to do in the first place. This worked quite well for simple systems but anything more complete than that ran into serious design problems.

No single software development fashion has solved the problems of poor adaptability. However, over the years, lessons have been learned and the most effective software developments incorporate ideas and lessons learned from each of the fashions over the last forty years. The currently most effective method of ensuring good adaptability is to use object oriented design. There has been a great deal written on this subject but the fundamental ideas are simple enough. In essence, object oriented software follows the natural law of objects in the real world. A cup for example, is an object that has a simple function. It acts as a container for liquids and facilitates drinking the liquid. The cup doesn't need to know anything about the liquid or the person drinking. It simply needs to be able contain the liquid and allow drinking. So it doesn't matter if you change the person drinking or if you change the liquid in the cup. You don't need to do anything to the cup.

FIGURE 13.1

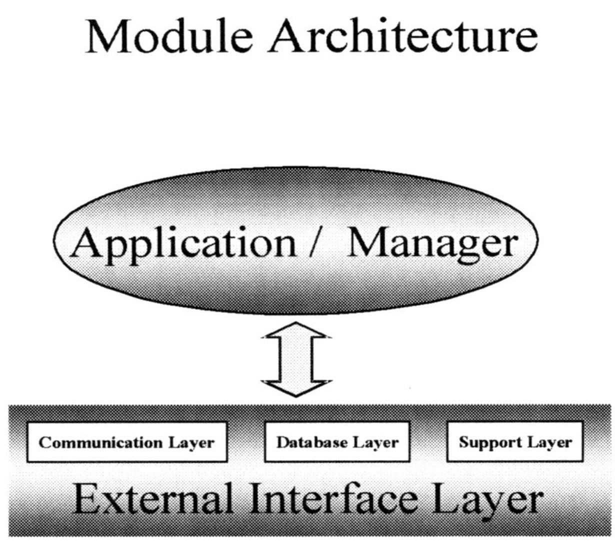

If systems are designed along the principals of object orientation then it is possible to greatly enhance the adaptability. Following the analogy further, you don't need to do anything to the cups, saucers and spoons in order to drink coffee instead of tea. This over simplifies the problem, in the real world of complex systems to support energy trading, there is more to it than that. Nevertheless object oriented designs, which have been rigorously followed through into the software package, are far superior in their adaptability than packages that do not have this principal enshrined in the design.

There are other major advantages too. In a package designed with object orientation, each object or component in the software package has its own distinct identity and its own distinct methods of interacting with other components. It means that new components can be added to the package as long as the new component knows how to communicate with its neighboring components. New components can be bought off the shelf and added to the package as long as the off-the-shelf component speaks the same language as the package. The necessary technology for standardized communication between objects is now available, for example COM by *Microsoft*.

How does object orientation (as a solution to the problem of poor adaptability) relate specifically to the world of energy trading and how does it link with the principle of flexibility discussed earlier? Using an example of flexibility in design that was examined earlier, we dissected a commodity transaction into its basic elements. The elements are:

- A quantity of commodity (this can be real physical commodity or it can be virtual or "notional"),
- A location (origination and delivery),
- A date (which includes time, possibly to the nearest minute)
- An amount of money (i.e. the price of that quantity of commodity and the currency denomination).
- A counterpart with whom the transaction is made

Flexibility is achieved by being able to design new types of transaction, which are made up of the basic building blocks. Take a swap transaction for example. A swap agreement is made with one counterpart. On a series of dates a series of quantities of a commodity are delivered to a location. In the case of a swap however the commodity may never be physically delivered. In this case the commodity quantity is virtual or "notional". The series of consignments could be variable. The price to be paid for each consignment may be fixed in advance or it may be variable, depending on some price index variable. The point though is that a swap is a combination with variants of the basic elements that make up any transaction.

Adaptability is achieved by designing the package to make use of object orientation. Each of the basic elements is treated within the software package as a separate object that has a relationship with its neighboring objects. Each object doesn't need to know how the other objects work it just needs to be able to do its own job and know how to communicate. In that way, new components and new variations can be added, building onto the complexity of the system but without increasing the risk of destabilizing the rest of the system. The behavior of existing components is not affected by the introduction of new components.

CONFIGURABILITY

Global energy markets require a global response from energy companies. Markets are merging and over-lapping both in terms of regional markets and in terms of energy commodities. In Europe for example, the market for electricity is relatively mature in the Nordic region but is still in the early stages of development in the German-speaking region. The gas market in Europe looks as though it will open up on the back of the emerging electricity market. In the UK the process is reversed. The UK gas market is relatively mature whereas the electricity market is just opening up. The New Electricity Trading Arrangements (NETA) is scheduled for implementation in October 2000.

Energy companies looking to participate in these markets need to consider a strategy that includes both gas and electricity, and one that also incorporates the different regional markets. It is essential that trading risk management systems can be configured to take account of the regional markets.

There are many ways in which a risk system could be configured. It depends entirely on the energy company in question on how this should be done. Therefore the risk management system chosen must be configurable to fit with the business not the other way round. Objected oriented design once again provides a solution.

The main business objects within a package for energy risk management are:

- **Trade Manager**—controls all aspects to do with trading, capturing deals, checking the validity of deals, preparing summarized information to be presented to the dealers etc.

- **Risk Manager**—controls all aspects of risk management, calculating the mark to market value of trades, keeping positions, calculating Value-at-Risk, running worst case scenarios etc.

- **Report Manager**—controls all aspects of report writing, running standard reports, composing new reports.

- **Curve Manager**—controls all market information and formation of forward price and volatility curves

FIGURE 13.2

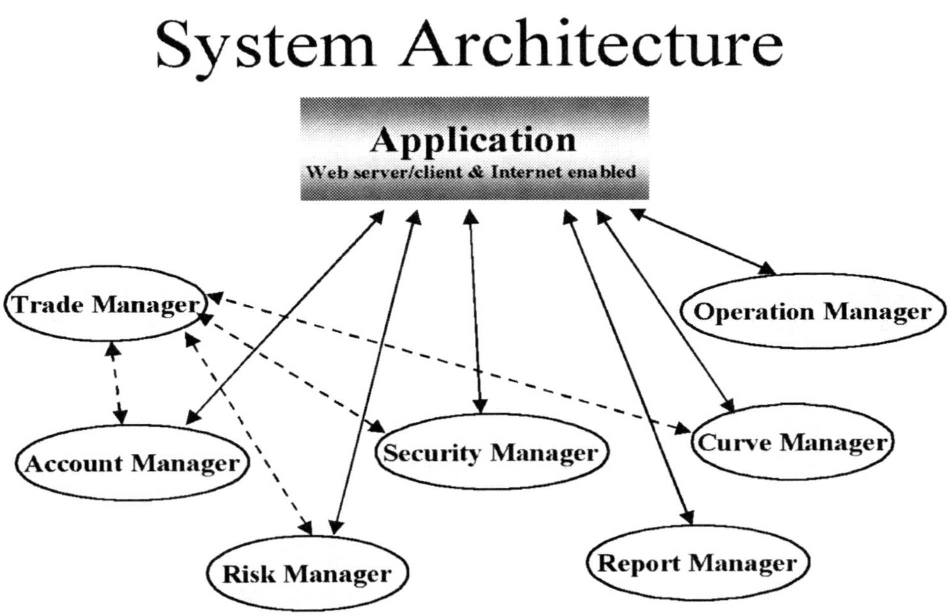

- **Security Manager**—controls all aspects of security, auditing and policing any activities.
- **Operations Manager**—controls all aspects of trade settlement and delivery of physical.

Each "manager" is a business object server and communicates both with end users of the package, the "clients," and with the back-end database server. This is tier client server architecture, providing enormous advantages in terms of configurability.

With a tiered architecture, as described above, it is possible to have a number of different trade managers each supporting a different trading desk. One trade manager could be serving an electricity trading desk, whilst another trade manager serves a gas trading desk. The trading locations of the two desks need not necessarily be in the same physical location.

Another advantage of this architecture is that the system can be distributed over a wide physical location. Consider the situation where an energy-trading organization has two physical locations trading two different commodities at the two locations. Group A trade electricity in region A and Group B trade gas in region B. Group A are the main trading group who also trade in region B. Group B does not trade in region A. The energy company wishes to configure their use of the risk management package so that Group A has its own trade manager but has access to forward curves for both localities. Group B has its own trade manager but only has access to forward curves for its own locality.

FIGURE 13.3

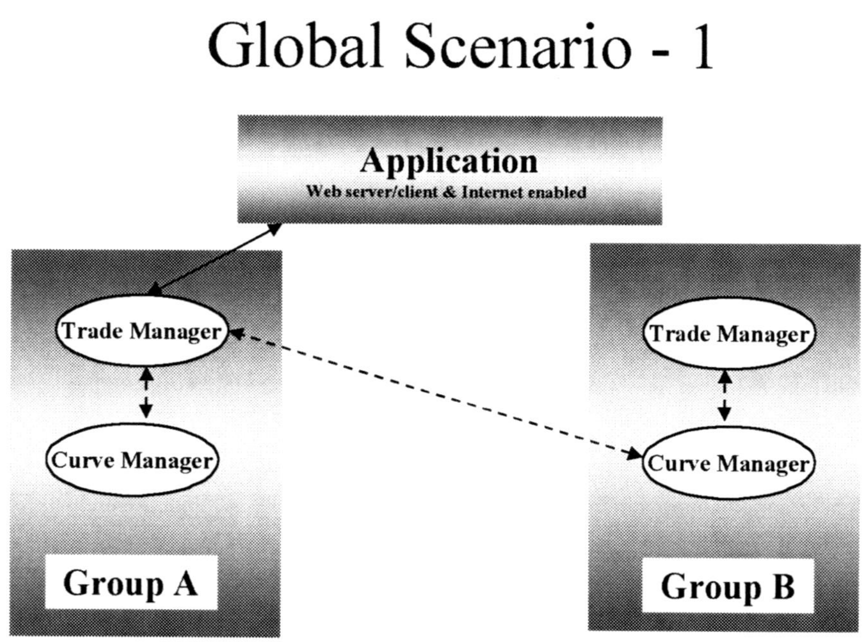

Another example could be that central risk management takes place in say, London, but that regional trading activity is carried out in New York and in Tokyo. The system needs to be configured such that trading can take place in the two regional trading offices but that central risk management takes place in the central office only. In this case the two regional offices have their own trade manager and curve manager but the central risk management office, in London, has a risk server.

This type of configurability needs to be built into the design of the pack-

FIGURE 13.4

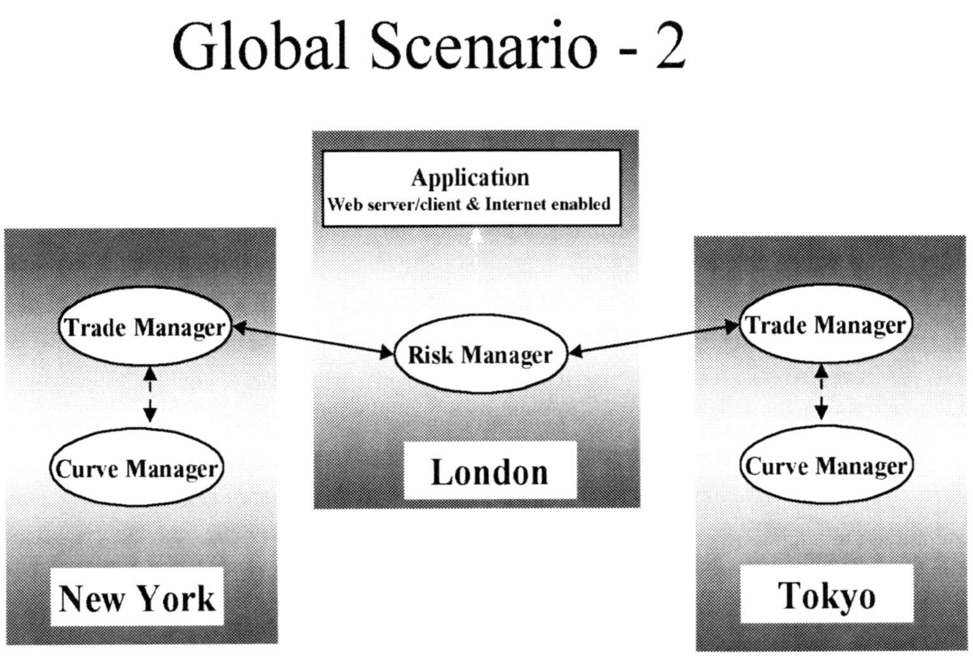

age. It cannot be incorporated as an after thought. Many of the problems faced by energy companies looking to expand into trading in different markets will be problems of lack of configurability of systems. The usual alternative is to replicate systems at each location and then to manually consolidate risk by extracting information and re-keying it into a risk database. This requires additional resources and is fraught with problems of double counting.

SCALABILITY

As the energy markets expand globally, energy companies will need to expand their trading operations to keep in line with the opportunities. Decision support systems and risk management systems need to display the same capability to expand usage in line business expansion. Once again this is not an issue to be taken lightly. Just as office space can become a limiting factor to the growth of a business so can capacity of systems. The problem with systems expansion is that you cannot necessarily take out temporary space in a nearby building. If system capacity limitations are reached it may be a case of starting again with a new system. Neither is it a sensible strategy to over purchase system capacity when current business needs do not warrant it. This would not only be adding unnecessary start up-cost but could prove to be completely unnecessary if markets do not grow in the way anticipated. The problem is that expansion of the energy markets is unpredictable.

The answer is to invest in a package solution that is scalable. For example, suppose that the initial requirements for an energy trading division may be to have three electricity traders, plus four administrators. Planning for twelve months time to have two trading locations with an additional five traders and four more administrators. The package must be able to expand usage to cover the additional site and accommodate the additional users without causing bottlenecks in the system performance.

Multi-tier architectures make this possible. Any number of new system users, "clients", can be added to the configuration without impacting performance or hitting any hard limitations. The configuration can be changed and fine-tuning can be carried out to optimize performance across the entire system.

The ability to use the Internet as a highly cost effective way of widening the distributed use of systems is becoming increasingly important. The architecture of the risk management package selected must be able to support use of the Internet. The advantage of using the Internet is that users of the risk management package can be located anywhere in the world that has Internet access. Low cost desktop computing allows the formation of "thin clients". The computing power required on the desktop need only be limited, sufficient only to be able to present the information required by the system user on their terminal. The real computing power can be located elsewhere on the Internet where the business-object managers are located. These objects perform the heavy computing tasks and can be located on suitably high performance computers communicating with the third tier to access database information. The data traffic can be kept low by configuring the use of the business object managers so that high band width is only required for communication between certain business objects and the database tier through a transaction processing layer (TPL).

FIGURE 13.5

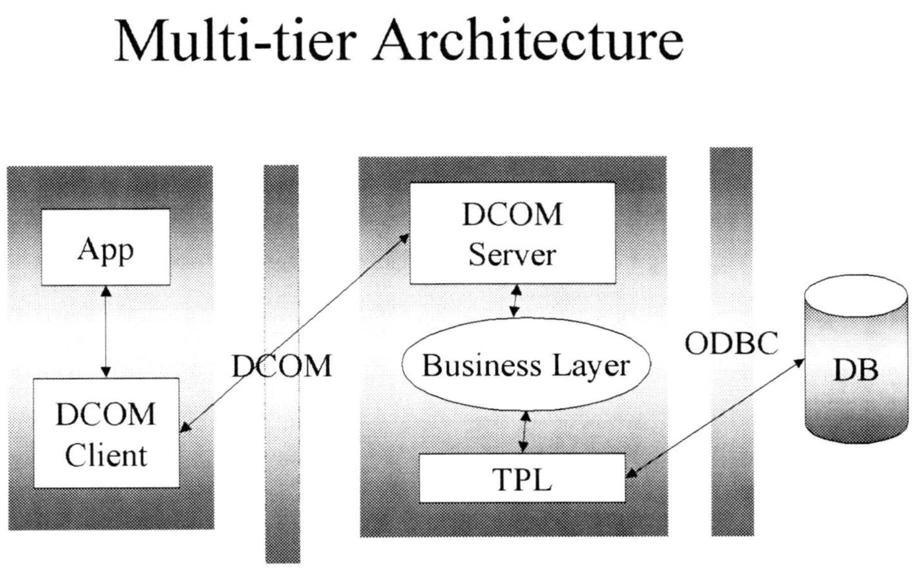

CONCLUSION

Two forces for change are bringing about the most radical upheaval in the energy business. The first is global deregulation of monopolistic energy supply and the second is the information technology revolution. This combination is producing a new market in which information technology will be the key. Interaction between physical networks for supply of energy, power grids, gas pipelines and virtual trading space, (the Internet, cyber space) are creating for the first time in human history a market place which contains an interaction of physical reality and virtual reality in a way that will fundamentally affect every consumer of energy. Within a short space of time intelligent machines will be automatically trading on behalf of individuals and organizations, switching energy suppliers on a micro-second timescale as prices for energy commodities fluctuate on the global markets. Intelligent air-conditioning systems, for example, will not only regulate room temperature in time with the weather but also in time with electricity pricing. Linking directly to automatic price feeds they will switch off or switch supplier when the market price gets too high.

Although this sounds like a vision of the future more akin to science fiction, it is becoming actual fact. Systems are being invented, created and installed which will make this a reality within a short space of time. Some energy companies will be left behind in the race and will cease to exist. Others will take up the challenge and become the new breed of energy company, able to juggle commodity production, consumption, and transportation in a complex market that requires microsecond-buying decisions.

In the coming months energy companies must decide to embrace the changes or seek the calmer waters of some easier market place, if they can. Between the old world of trading and the new lies the uncharted waters of implementing decision support and risk management systems which will provide the information necessary to navigate the new world safely.

Not all risk management systems currently on the market are up to the job. If they are to perform the task of enabling energy companies to manage the risks of trading in the new paradigm then certain pre-requisites are needed. The risk management systems must be designed to deal with the process of change that is occurring.

They must be designed to be flexible—flexibility means being able to react to innovations in the way that the market trades, it means being able to deal with multiple commodities, it means being able to deal with multiple markets and regions.

They must be designed to be adaptable, meaning being able to add new modules quickly so as to be able to take advantage of new developments in the market, and to reduce exposure to the risk of not being ready. It means being able to implement change without destabilizing the remainder of the system. It means being able to make modifications without having to change the entire system.

They must be designed to be configurable, meaning means being able to optimize the use of the risk management system so that it reflects the actual use made of the system by the business. It means having a risk management system that reflects the business and not a business that reflects the risk management package.

And they must be designed to be scalable, meaning that the system grows as and when the organizations use of the system grows. It means that bottlenecks are not created and performance does not diminish as usage increases.

Two techniques of system design can be brought to bear. The first is the use of object oriented design techniques and the second is the use of a multi-tier architecture incorporating Internet and Web enabled technologies.

CHAPTER 14

Wellhead to Wires: Energy Convergence

By Peter C. Fusaro and Jeremy Wilcox

CONSOLIDATION NOW UNDERWAY

The new energy market that is emerging is being driven by several mega trends. These market drivers include rapid technology change, growing worldwide environmental awareness, market deregulation and privatization, globalization and most importantly financial innovation and securitization. The new energy industry is larger companies involved in oil, gas and power, reintegrated from their upstream production, refining, generation and distribution services to its customers. It is an industry with more financial risks due to the deregulation and market-driven economics of this age.

The new industry is entering a period of maximum market consolidation with the BP Amoco Arco, Exxon Mobil, and TotalFina Elf mergers leading the way. Size matters in global energy markets. A globally, diversified portfolio of assets is the new game in town. Managing those assets becomes paramount. The new financial tools of energy trading make energy trading a survival skill for the industry.

The new technologies that are reshaping the industry are coming from two faces of the industry. One is the technological change in extracting, transporting, refining, and selling oil, gas and power through 3-D seismic, horizontal drilling, combined cycle gas turbines, fuel cell and micro turbines. The other end is that of information technological. IT creates value along the supply chain. This information becomes a financial commodity that can be traded. That is why we now see the new markets of bandwidth trading.

The realization is that value creation is the power generation sector and in providing new financial and bundled services to its customer base. This includes telecommunications, cable, and energy services to its customers.

The new emerging financial markets for weather, bandwidth, and emissions are global markets. They are harbingers of change, which manifests itself over every facet of the energy industry. Newer markets for LNG hedging have already emerged. The market for retail financial products is being to be developed due to deregulation at the retail customer level i.e. fixed price bills and price caps for energy bills.

More importantly, the industry will become a major adapter to Internet technology for trading. Our report, **Electronic Energy Trading**, has had to be updated every four to six months to keep abreast of all the electronic platforms launched and under development. At the present time, most of these platforms are North American-focused due to the heavy market penetration of the Internet there. But Europe, Asia and Latin American are sure to follow rapidly over the next two to three years. This accelerates the change process and move to market of new trading platforms. These will all be in cyberspace even from the existing energy exchanges, NYMEX and IPE, which realize the need to be web-oriented. What this means for financial markets is there will be rapid adaptation of financial energy trading platforms for the Internet.

The Petroleum Electronic Exchange (PEPEX) is a perfect example of the changing dynamics of energy trading. Traditionally, state-owned oil companies tendered their production of crude oil and petroleum products using an antiquated system of trading confirmed by TELEX, another old telecommunications technology. On March 27, 2000, tendering entered the age of the Internet and electronic trading. This acceleration of the dinosaurs of the oil industry into the Internet age signals that oil trading will move all players to the industry.

As clearly has been shown with the oil price move from $12 dollars per barrel to $34 per barrel on NYMEX from the end of 1998 to early 2000. Oil price volatility still exists. Oil trading on futures exchanges is still growing twenty-two years after the launch of the first successful energy futures contract in 1978. Several electronic trading platforms are being launched for crude oil, petroleum products, natural gas liquids (NGLs), natural gas, electric power and petrochemicals. These Internet-based exchanges are broadening the energy trading market both cross borders and globally.

Financial innovation continues to penetrate energy markets. In 1999, over $2 billion of LNG cargoes were hedged. This had never been done before, but the new market dynamics on LNG including smaller supply trains, over supply, and the development of a global spot made it possible to hedge LNG for producers and consumers.

The self generation space of cogeneration, fuel cells and micro-turbines will also develop into a e-commerce and trading platform as market liberalization, new technologies, and new market needs drive creativity. Energy efficiency and renewable fuels add to this new stew of supply choices, customer driven markets, and more financial risk. Moreover, this change process is disruptive, discontinuous, nonlinear and accelerating. It is a time of continuous technology improvement and rapid adoption of these newer and more efficient technologies. Energy risk management now becomes a core competency to compete in global markets since risk is now pervasive.

The dawn of the Internet age is bringing with it a quick transference of new ideas, information and technologies. It is also bringing with in more risks. These are price, market, event, counterpart, legal, technology and regulatory. But it seems that first mover advantage is the new adage in these times.

The new world of energy trading is bringing with it relentless change to 24 hour markets, seven days per week, and based on the Internet. The rapidity of market participants, first in North America, but all over world demonstrate that financial innovation, financial engineering and energy trading is actually just getting started. The best is yet to come as cross commodity arbitrage of power with bandwidth and emissions with fuels show that there is untapped potential for more financial engineering in this sector.

Deregulation and globalization of the energy markets are bringing with it market-based solutions such that protectionism is dying. It inevitably brings more risk. Price volatility for oil, gas and electric power will always exist, but price spikes will be shorter lived. Moreover, financial market maturation will take the next 20 years. The new commodity markets of emissions, weather, and bandwidth promise to change investment decisions and will have unforeseen consequences.

The more innovative work is also beginning in energy project finance with the imbedding of emissions derivatives in new power station financing. Energy securitization and monetization promise to expand rapidly in the new world of energy.

What's already emerging is that the Over-the-Counter (OTC) markets for energy are moving to the Internet and are being created quickly and adapting instantaneously. Electronic exchanges are truly global and virtual and are multi-commodity. The movement of energy business-to-business (B2B) commerce is just beginning with fuel and equipment procurement, billing and meter reading moving energy trading and outsourcing solutions starting to shift to the web.

CONSTANT PROCESS

Energy industry change is now being driven by a multiplicity of factors impacting its structure dramatically. Continuous oil and gas price volatility, efficiency gains through the wider adoption and application of information technology, continued commoditization of international and regional energy markets, environmental market drivers, globalization driven by deregulation and privatization, and industry consolidation are transforming and recreating a reintegrated energy industry. This new industry is market-driven, supranational, and continuing to metamorphosize. The rate of change cannot be underemphasized since it is discontinuous, nonlinear, accelerating, and constant. The e-commerce dimension to this equation only exacerbates these underlying phenomena.

The question becomes why is the industry changing now? The answers are that since the multiple market drivers that cannot be adequately modeled, quantified or predicted, something new is occurring in this space. The old modalities including thinking must be jettisoned for the new to be created. This is almost analogous to Schumpeter's creative destruction model, but I liken it more to Jungian psychology in that it is the unconscious mind or dreams that is, in effect, creating the future. The tangible and quantifiable and discernable trends are very simple to understand: globalization, deregulation, privatization, technology innovation and transference, financial innovation and securitization, worldwide environmental awareness and consciousness, and change. The indiscernible and aggregating change components cannot be measured. Change is not the mantra but the key component of our times. It has become the fabric of the energy industry. It is a disruptive process. Like it or not!

The rapid transference of information, ideas and technology has only just begun. These times have been compared to the Gutenberg printing press and the Reformation. That is too European and Western centric. The reality is that change is constant, global, and rapidly transferable wherever it comes from. It is actually a more Eastern philosophy. Change is constant, all encompassing and immutable. The Internet should be looked at as an organic matrix where ideas are transferred, synthesized, revised and adopted globally by a collective consciousness.

Getting back to the assault on the consciousness of the energy industry, the new industry is reintegrating. The new markets for oil, gas, and power are global. They are greener and more energy efficient. They are driven by customer choice. They are technologically integrated and embrace new technologies. They are market driven not solely created by government fiat. The value creation is all along the energy supply chain from the production to the end-user behind the meter. Information and ideas now drive the new emerging e-commerce energy industry, from the wellhead of oil and gas production, through the value chain of pipelines, refineries and jobbers, to the burner tip and beyond.

The aggregate of smart meters and buildings, advanced SCADA systems, more efficient turbines, lighting and HVAC systems, and real time modalities for energy information from production through consumption are manifested as a consolidating energy industry providing for a much smaller work force and a much larger global portfolio in oil, gas and power. More importantly, it will all be accomplished at lower prices as all facets of the industry become commoditized.

Harbingers of these changes are everywhere. The rapid market awareness by financial markets recently of fuel cell and micro turbine technology illustrate this point with the soaring stock valuations of many vendors some of which will not have commercial products for several years. Other emerging technologies, such gas to liquids, are coming under a similar financial spotlight. The development of LNG spot markets in effect creating a global gas market with diversified supply choices, and the ability to not follow the traditional process of technology adoption and industry development. The advent of renewables resources as a commercial venture will be the next wave of market adoption and financial interest. Leapfrogging technological opportunities in the energy production and usage are now myriad.

These markets have more risks. Not just price risks, but supply event, technology, credit, and regulatory risk to name a few. Therefore, energy risk management and financial engineering skills become a core competency of energy companies. In fact, the skill set shifts from engineering to commercial and financial. Already, in energy project finance for power stations, we are seeing the embedding of environmental emissions derivatives. These factors taken in isolation may not seem important, but taken as an aggregated portfolio of change can not be underestimated as rapid adaptation of technology is a given with the information superhighway being created before everyone's eyes.

While there are of course unique characteristics of each country and market, the key commonality is the rapid dissemination of ideas and information. The global integration of the energy markets are key components of the convergence of the multicommodity energy trading markets, new technologies, energy efficiency, overriding environmental imperatives, and a transition to natural gas shorter term are all manifestations of this global reintegration process.

GAS AND ENVIRONMENT PICKING UP MOMENTUM:

The new energy world is a greener and gas driven world. Global gas reserves are well diversified and a global LNG spot market is rapidly evolving. The doubling of gas consumption projected on Table 1 are fundamentally altering how oil companies appreciate the added value of natural gas which is becoming the transitional fuel for power generation as growing global environmental awareness on carbon reduction will move the world fuels platform to cleaner technologies as an ongoing means of conducting business. Gas trading and ownership of gas-fired power stations can add further value to the commodity through a range of marketing and arbitrage opportunities. The following table only begins to indicate the rapid transition to natural gas usage globally. These volumes may be underestimated.

TABLE 14.1

GLOBAL NATURAL GAS CONSUMPTION OUTLOOK

Region	1997	2000	2010	2020
		(Billion cubic feet)		
Asia Pacific	8,901	9,171	11,740	16,243
Middle East	5,660	6,367	10,371	16,894
North America	26,129	27,232	35,227	43,381
Latin America	2,930	3,382	6,349	13,086
Western Europe	14,263	15,360	20,642	30,263
Eastern Europe/FSU	22,028	22,696	27,396	35,069
Africa	1,844	2,015	3,129	5,603
World Total	81,755	86,223	114,954	160,539

Source: 1997: US DOE's EIA, Forecast from Enron Corp

Global environmental awareness now drives capital investment and business decisions. The export of pollution to developing countries is now passé. The global media brings with it global environmental consciousness and awareness. It will influence energy industry investment more than anyone realizes. The increased concerns on global warming and greenhouse gases are about to unleash a tidal wave of new and more benign technologies on this industry. Natural gas usage will be used as a transitional fuel to meet environmental standards over the next two decades. Emissions trading schemes are but one element that will lead to technology and income transference to the developing world where energy demand is now nascent but is growing. The movement on global warming is now a decade late. The financial markets will accelerate the adoption of more benign technology and accelerate its commercialization.

Two of the major issues dominating the European energy market are gas competition and emissions trading. A major catalyst to gas competition will be the implementation of the European Union's Directive on gas competition in August 2000. Similar to that of last year's electricity directive it aims to establish a competitive European gas market and will be one of the final steps to form a single European energy market. The directive sets out the rules for the transmission, distribution, supply and storage of natural gas. The comprehensive scope of the directive addresses the organization and functioning of the sector, market access, system operation and the criteria and procedures for granting of authorizations for transmission, distribution, supply and storage.

As with electricity, market opening will be achieved through a combination of eligibility criteria for customers to have access to competitive supplies. Initially, the threshold for access to competitive supplies will be limited to all gas-fired generators and other final customers consuming in excess of 25 mcm of gas per year. This threshold will reduce to 15 mcm in 2003 and by a further 5 mcm in 2008. In percentage terms, 20% of the market will be open to competition from this year, 28% in 2003 and 33% in 2008.

In tandem with electricity competition the gas thresholds are expected to be exceeded. Analysis of individual market competition indicate that up to 34% of the market will be open this year, 36% in 2003 and 42% in 2008. Also, as in electricity the market opening will be skewed with some EU markets more competitive than others. The UK is already 100% open, Spain has 46% competition already, the Netherlands will be 45% by the end of this year, Belgium opened 47% of its market last year and finally Ireland already allows third party access to 75% of its market.

The obvious benefit of linking gas and electricity competition is in the creation of a more convergent energy market, which is discussed in more detail below. The UK market is a prime example of the convergent opportunities of gas electric from generation through to domestic (retail) households. But the path to gas competition will likely be problematic. Uneven competition, with some areas more competitive than others, will impede trading development and, as with electricity, the EU might be forced to take legal action against those member countries that delay competition.

A similar problematic path is faced by emissions trading. But the problem is more a factor of how to introduce it rather than disagreements over its benefits. The European Union is strongly supportive of the Kyoto Protocol and a number of EU member countries have identified the benefits of emission trading to reduce pollution levels. The issue then is not whether a trading program is introduced, but how it is introduced.

At the center of the debate is how to allocate emission permits—grand fathering, benchmarking or auction. The older generation plants support the process of grand fathering, as it is more beneficial to them. Benchmarking is a positive route forward but is dependent on accurate historical emission data. Finally auctioning is an unbiased approach but raises the question of how the value of the emission allowances is set. As in most situations like this a compromise approach is the best way forward. Such an approach would involve the initial phase of permit trading—say five years—based on a hybrid grand-fathering / benchmarking approach to setting levels followed by auctioning thereafter.

An alternative approach to 'standard' emission trading has been proposed in Sweden. The concern expressed in Sweden is that standard emission programs tend to support investment in high emission generation plant. Sweden, by virtue of its high hydro and nuclear generation mix has minimal emissions and it wants to incentivize development of this type of plant. The solution may be Climate Change Certificates, which allocate a greater number of permits for green generation such as hydro than for less green generation such as coal and gas. The end game though remains the same. Reduced emissions through trading.

This widespread support of emission trading should lead to some EU countries setting up trading platforms before the end of the year. Certainly it makes sense for these platforms to be developed in tandem with increased gas competition as Europe addresses its environmental conscience. But as with other markets it remains to be seen whether Europe can develop a single unified market approach to emission trading.

THEORY OF CONVERGENCE:

Market convergence is a natural conclusion of competition and consolidation. It is also related to the existence of correlation and causation between similar, but different, market products. As such we refer to convergence as the 'Three C Strategy'. We address both routes to convergence below.

The onset of market liberalization (also referred to as deregulation) has been the major catalyst for change in the global energy market over the past twenty years. Until the last decade competition in the energy market was restricted to the oil sector, but the progressive, and unfinished, liberalization of the gas and electricity markets is proving to be the primary catalyst for fundamental change. The starting points were gas deregulation in the US began in 1978 and completed in 1993 and electricity liberalization in the UK begun in 1989. The momentum generated by these is reshaping the structure of the global energy market as well as opening up new opportunities in emissions, weather and bandwidth / telecommunications. The ultimate end game will be a convergent energy market spanning oil, gas, electricity, weather, emissions and telecommunications and controlled by a small number of international corporations.

While some way off the seeds of a convergent energy marketplace have already been sewn. With competition increasing at a rapid pace energy utilities are addressing consolidation strategies to both increase economies of scale and their competitive edge. The plethora of electric utilities, for example, is gradually being reduced as the competition stakes are increased. Part of this consolidation is being driven by regulatory policy as electric markets move from wholesale to retail access. In the UK, the Office of Gas and Electricity Markets (Ofgem) has levied strict price controls aimed to reducing consumer costs. As a result Utilities are being forced to cut costs in order to comply, the solution being mergers to achieve greater economies of scale.

A similar development can be seen in Germany with the agreed merger of RWE and VEW to form the continents largest electricity supply company. Elsewhere in Europe and the US there have been similar mergers and acquisitions, driven by competition. In a way it is ironic that in setting out a policy for competition the number of market participants are contracting. This raises the question of monopoly and oligopoly practices emerging. But then how many participants constitute an oligopoly? In the UK gas / electric market it is widely accepted that the market can be efficient and competitive with just five supply companies—a reduction of over 60%. In Europe as a whole the 5,000 plus gas / electric utilities will contract to less than 1,000 by 2004. Similar levels of consolidation will be reflected in North America and Australasia.

But what of the link between consolidation and convergence? This final step in the three C strategy is derived from the forces of correlation and causation. Correlation implies that price movement in two markets is linked. But for correlation to be effective there has to be causation between these markets. That is to say that the price movement in one market causes the price movement in another. A well-used example to illustrate this point is the strong correlation between fuel oil prices and that of pork bellies. While historical price movement between the two is correlated the causes of price movement are not.

This is not the case in the energy market and as such becomes a driver for convergence. The correlation between oil and gas is well documented. Not only is gas a by-product of crude oil production but also before the gas markets were effectively commoditized gas was priced based on fuel price escalators. Customers with interruptible gas supply contracts took fuel oil as the heating alternative with gas supply contracts priced out based on fuel oil price escalators.

The next strong link is between gas and electricity. Natural gas is increasing its share of the generation market and electricity suppliers are increasingly taking physical gas assets. The link is two-fold. From the production side generating companies with gas assets can assess the value of gas based on the economies of converting gas to electricity through spark spreads. If the market price of gas is high and the cost of converting this gas will generate electricity priced above the market then it makes economic sense for the producer to sell the gas to end users and buy in wholesale electricity to meet its supply commitments. Conversely, if the gas price is low and the cost of converting the gas to electricity leads to a competitive electricity price then the gas will be used to generate electricity. This arbitrage opportunity is obviously dependent on imperfections between the two markets and will gradually diminish as gas assumes a greater percentage of electricity generation.

The same argument can be applied to coal and it is the liberalization and competition in the electricity markets that is acting as the catalyst to the commoditization of coal. But the link between coal and electricity goes further. Of prime importance to the global community are emissions and coal fired generating plant is the greatest emitter of emissions, particularly sulfur oxides. Also, with coal in the early stages of commoditization the need to standardize and accurately quantify contracts becomes important. This process has been assisted by the development of trading sulfur allowances in the US.

If a trader knows the price of sulfur oxide allowances he can attach a dollar value to the sulfur level of the coal, thereby accurately pricing out the coal contract. Previously traders would use bartering methods to value the greater, or lower, sulfur content of the coal against the required contract specification. But by employing emissions values the pricing of coal contracts is improved. It can be argued that the link between coal and emissions, and therefore electricity, enables a BTU value to be attached to emissions.

Two other markets complete the convergent energy spectrum—weather and bandwidth. The link between weather and electricity / gas is obvious. Lower temperatures increase the heating requirement and therefore have a direct impact on gas / electricity prices. Conversely higher temperatures increase the cooling requirement through air conditioning, thereby having an impact on electricity prices.

Unlike electricity, historical weather data goes back over 100 years. Given the correlation between weather (primarily temperature) and electricity it is possible to generate forward electricity price curves based on weather data. Again, as for emissions, weather can be assigned a BTU value and traded as such.

The last addition to the convergent energy market is bandwidth. Here the link is with the electricity market as the peak and off peak prices are correlated between the two. Electricity and telecommunication usage is greatest in the morning and early evening periods and is minimal overnight. But the greatest value in terms of convergent opportunities is in the retail sector. The customer base of gas, electricity and telecommunications is similar. In addition telecommunications have a stronger branding than gas or electricity. These factors are combining to create the genesis of the mega, or convergent, utility on both sides of the Atlantic.

From this analysis the convergence and correlation between oil, gas, electricity, weather, emissions and bandwidth is apparent. And this theory is already being put into practice. Large oil companies such as BP Amoco are investing in electricity markets while traditional electricity supply companies are converging into oil. And both sides are reaping the opportunities of weather, emissions and bandwidth.

CONCLUSION:

Worldwide deregulation and energy industry consolidation is now converging into a global portfolio of energy companies with assets from oil and gas production through the value chain to electric power generation. In the past, there have been several waves of energy industry restructuring driven primarily by oil price volatility. The new focus is now changing toward global competition in energy market that will supercede the independent elements of energy production, refining, and distribution systems to encompass the new market drivers of technology, environment, commoditization, and global competition. It is leading to lower margins, alliances, mergers and acquisitions. The need to enhance value is in the midstream gas/electric marketplace, which is leading oil companies to own generation assets. The next step would be to create the energy solution provider through an energy services company, which would offer gas, electricity, water, telecommunications as well as engineering services to maximize value for its customers and deepen those relationships.

The new energy industry that is emerging is first and foremost technology driven as well customer driven. Customers will reshape the industry at all levels of the supply chain. Brand awareness will diminish as commodity markets usurp traditional price-driven supply relationships. What will be needed is true customer service driven by customer choice initiatives throughout the world. Incumbency advantages will diminish over time, if energy companies do not deliver customized solutions and services to their clients.

The new emerging energy industry is reintegrated as companies recognize the value of their assets across the supply and value chains from production to end-use. The ultimate consolidation of the industry is larger companies with a global portfolio of producing, refining, and generation assets. Companies still have to produce goods and services even in the

Internet economy.

The new energy industry is now beginning this process of reintegration where it has a global portfolio of production refining and power generation assets. This movement toward globalization inevitably concentrates market power into the hands of fewer players with many companies reduced to niche markets. The emerging industry is driven by capital markets and technological needs to become the full energy service provider where feasible and those functions outsourced where necessary.

The new energy equation is to securitize oil and gas production throughout the value chain to customers. These changes are now manifesting themselves throughout the world as deregulation, commoditization, and competition drive the energy industry into a new configured provider of services as well as goods.

The nature of these changes is compelling energy companies to adapt more quickly than ever before. First movers have advantages in times of rapid change as is evident today. Flexibility and quicker decision making drives these choices as adaptive organizations must evolve in order to survive. Mistakes are inevitable in this environment of experimentation. The technological imperatives impact exploration and production, manufacturing, power generation and distribution of energy goods and services. The Internet economy, which is just starting its evolution into the business-to-business space only exacerbates the rapid adoption of new business processes and disseminates that information globally with minimal barriers to entry. Commodity markets provide the price transparency necessary to globalize oil, gas and power. They move the control of assets from producers to their customers since supply side options open up.

This is the age of information and technology. The pace of change is accelerating. First movers win. We are now in the beginning of the consolidation phase of the market as global competitors are developing their wellhead to wires strategy. There are no models, data, or templates to predict this recreating future. There are just new ideas, experimentation, failure, and flexibility. It is truly a new age for the energy industry or is it the New Age Energy Industry?

Glossary of Energy Risk Management Terms

American Option—An option which can be exercised by the buyer at any time during its life.

Arbitrage—The simultaneous purchase and sale of similar or identical commodities in two markets.

Asian Option—An option which can be exercised only at expiration, based on the difference between the strike price and the average of daily spot prices over the life of the option. Also called average price options.

Backwardation—Market situation in which prices are progressively lower in distant delivery months. Opposite of contango.

Basis—The differential that exists at any time between the futures or forward price for a given commodity and the comparable cash or spot price for the commodity. Basis can reflect different time periods, product qualities, or locations.

Basis Risk—The uncertainty as to whether the cash-futures spread will widen to narrow between the time a hedge position is implemented and liquidated.

Bid/Ask—A measure of market liquidity. The bid is the price level at which buyers are willing to buy and the ask is the price level at which sellers are willing to sell.

Book—The total of all physical, futures and OTC derivatives positions held by a trader or company (includes documentation).

Broker—In futures, the person who executes the buy and sell orders of a customer in return for a commission or fee. In the OTC markets, the person who introduces counterparties and arranges the transaction charging a fee for this service. Brokers never take a position in the market.

Calendar Spread—An option position created by selling one call and buying another with a longer expiration at the same strike price. Also called Time Spread.

Call Option—An option that gives the buyer the right (but not the obligation) to enter into a long futures position at a predetermined strike price, and obligates the seller to enter into a short futures position at that price, should the option be exercised. See put option.

Cap—A contract between a buyer and seller in which the buyer is assured he will have to pay more than a maximum price. See Floor.

Carrying Charge—The total of storing a physical commodity including storage, insurance, interest and opportunity costs.

CFD—A contract for difference which is basically a price swap and usually applied to the short term Brent crude oil swaps market.

CIF—Cost, insurance and freight.

Clearing Member—Members of an exchange who accept responsibility for all trades cleared through them and share secondary responsibility for the liquidity of the exchange's clearing operation. Clearing members must meet minimum capital requirements.

Clearing House—An exchange-associated body charged with insuring (guaranteeing) the financial integrity of each trade. Orders are cleared by the clearinghouse acting as a buyer to all sellers and seller to all buyers. The clearinghouse stands behind all trades made on the exchange.

Close—The period at the end of a trading session and is also called the settlement price of that commodity.

Collar—Options strategy designed to minimize upfront costs of a cap or floor though the sale of a cap or floor.

Contango—Market situation in which prices are progressively higher in succeeding months than in the nearest delivery month. Opposite of backwardation.

Counterparty—The person or institution standing on the opposite side of a transaction.

Credit Risk—The risk of default by either counterparty in a transaction.

Crack Spread—An intermarket spread where futures are bought and sold to mimic the refining of crude oil into petroleum products.

Delivery Month—The month in a given contract for delivery of the physical commodity.

Derivatives—Financial instrument derived from the underlying commodity including forward, futures, swaps and options.

European Option—An option which can be exercised by the buyer only at expiration.

Exotics—A term used to refer to more structured over-the-counter instruments, such as swaps with a range of maturities, volumes or fixed prices, transactions that incorporate different commodities, or complex combinations of options.

Floor— A contract between a buyer and a seller in which the seller is assured he will receive at least a minimum price. See cap.

FOB—Free on Board. A transaction in which the seller provides a commodity at an agreed upon price, at a specific loading point within a specific period of time. The buyer must arrange for transportation and insurance for delivery.

Forward—Standardized contract for the purchase or sale of a commodity which is traded for future delivery not under the provisions of an established exchange.

Fungibility—Characteristic of products or instruments that can be comingled for trading, shipment, storage or consumption. Only fungible goods can be traded as commodities.

Futures—Standardized contract for the purchase or sale of a commodity for future delivery under the provisions of exchange regulations.

Hedging—In the futures or OTC markets, a simultaneous initiation of equal and opposite positions in the cash and futures markets. Hedging is employed as a form of financial protection against adverse price moves in the cash market. Opposite of speculation.

In-the-Money—An option which has intrinsic value at expiration. Opposite of Out-of-the- Money.

Initial Margin—Margin posted when a futures position is initiated.

Intrinsic Value—The value of an option if it were to expire immediately. Cannot be less than zero.

ISDA—International Swaps Dealers Association. A trade association, primarily of banks, that promotes the use of all derivative instruments. ISDA has its own master swaps agreement that is heavily used by the energy trade.

Liquidity—A characteristic of a market where there is a high level of trading activity.

Local—A commodity or options principal and exchange member who buys and sells for his own account on the floor of the exchange.

Long Position—In the futures market, the position of a contract buyer whose purchase obligates him to accept delivery unless he liquidates his contract with an offsetting sale. In the forward market, a long position obligates a buyer to accept delivery unless a book-out agreement is subsequently made.

Margin—Funds or good faith deposits posted during the trading life of a futures contract to guarantee fulfillment of contract obligations.

Mark-to-Market—Daily adjustment of open positions to reflect profits and losses resulting from price movements that occurred during the last trading session.

Market Maker—A dealer who consistently quotes bid and offer prices for a commodity. Can be an energy company.

Moving Average—Technical analysis term which clearly signals any change in the trendline.

Net Position—The difference between an entity's open long contracts and open short positions in any one commodity.

Notional—The underlying principal of either an exchange traded or OTC transaction.

Offer—A motion to sell a futures, forward, physical or options contract at a specific time.

Offset—A transaction that liquidates or closes out an open contract position.

Open—The period at the beginning of the trading session on a commodities exchange.

Open Interest—The number of futures contracts on an exchange which remain to be settled.

Option—The instrument that gives the holder the right to buy or sell the underlying commodity at a given price or at a specific date.

Out-of-the-Money—An option which has no intrinsic value at expiration. See In-the-Money.

OTC (Over-the-Counter)—Purchase and sale of financial instruments not conducted on an organized exchange.

Paper Barrels—Term used to designate non-physical oil markets including futures, forward, swaps and options.

Position Taking—The action of commercial participants who use the futures market as an alternative cash market rather than as a hedging vehicle.

Premium—The price paid by the option buyer to the option seller.

Price Signalling—Advanced publication of prices prior to their effective date for the purpose of encouraging competitors to make similar price changes.

Put Option—An option that gives the buyers the right (but not the obligation) to enter into a short futures position at a predetermined strike price and obligates the seller to assume along futures position, should the option be exercised. See Call Option.

Short Position—In the futures market, the position of a contract seller whose sale obligates him to deliver the commodity unless he liquidates his contract by an offsetting purchase. See long position.

Speculation—The opposite of hedging in which the speculator holds no offsetting cash market position and deliberately incurs price risk in order to reap potential rewards.

Spot—A one-time open market transaction, where a commodity is purchased "on the spot" at a current market price.

Spot Month—The futures contract held closest to maturity.

Spread—The simultaneous purchase of one futures or forward contract and the sale of a different futures or forward contract. Also refers to futures/forward contract purchase in one market and a simultaneous sale of the same commodity in some other market.

Straddle—The purchase or sale of both a put or call having the same strike and expiration date.

Strike Price—The predetermined price level at which the exercise of an option takes place.

Swap—Customized contractual agreement between two parties to exchange interest payments, typically a fixed rate payment for floating rate payment. No physical commodity exchange takes place.

Swaption—A option of a swap.

Technical Analysis—Examination of patterns of futures price changes, rates of change, and changes in trading volume and open interest, often by charting in order to predict and profit from such trends.

Time Value—Part of the option premium which reflects the excess over intrinsic value.

Trend—Price activity in markets in a particular direction, and is characterized by higher highs and higher lows.

Variation Margin—Margin paid or collected in order to maintain a minimum level based on daily fluctuations in contract value.

Volume—The number of transaction occurring on an exchange during a specified period of time.

About Global Change Associates Inc.

Global Change Associates is an international energy and environment consulting firm with particular expertise in e-commerce, risk management, market regulation and restructuring, and competitive market intelligence. The company was established in New York in 1991 to service the North and South American energy markets. A European office opened in 1999.

The company provides customized solutions to clients in the oil and gas, electric power, energy services, emissions, information technology, and software industries, as well as to the investment banking sector and information providers. Its market intelligence global network for energy information is second to none due to the "high tech, high touch" philosophy of its principals.

Since 1999, the company has focussed on energy e-commerce, producing the first comprehensive study on global energy e-commerce from trading through procurement. The study—**Electronic Energy Trading**—was last updated in April 2000 with a third version due November 2000. Global Change Associates is now recognized as the leading independent authority on energy e-commerce.

For more comprehensive information, including online ordering of all the company's products, visit the company's world wide web site at www.globalchange.com.

ORDER FORM

To order a copy of **Energy Derivatives: Trading Emerging Markets**, either purchase direct from the web site (www.global-change.com) or fax the order form to Whitehurst & Clark: 732-225-1562.

Name: ____

Company: ____

Position: ____

Address: ____

____

Tel.: ______________________ Fax: ______________________

E-mail: ____

Please send me _________ copies of **Energy Derivatives: Trading Emerging Markets** at $85.00 per copy; 2 or more copies at $75 per copy (plus $5.00 US shipping; $20.00 foreign shipping).

- Please invoice me $__________
- I enclose a check for $_________ made payable to **Global Change Associates**
- Please debit my ❑ American Express for $_________

Card Number: ______________________ Expiration Date:____________

Signature: ___________________________ Date: _____________

Printed in the United Kingdom
by Lightning Source UK Ltd.
133789UK00001B/73-74/A

9 780974 539126